Management of Municipal Solid Waste

Management of Municipal Solid Waste

T.V. Ramachandra
Centre for Ecological Sciences
Indian Institute of Science, Bangalore

Commonwealth of Learning, Canada
Centre for Ecological Sciences, Indian Institute of Science, India
Karnataka Environment Research Foundation, India

First reprint 2009; Second reprint 2011; Third reprint 2013
Fourth reprint 2014; Fifth reprint 2014; Sixth reprint 2015
Seventh reprint 2016; Eighth reprint 2016; Ninth reprint 2017
Tenth reprint 2018, Eleventh reprint 2018, Twelth reprint 2020

Management of Municipal Solid Waste

Dr. T.V. Ramachandra
Instructional Designer and Editor: Dr. Krishnapillai Murugan

Production Assistant: Mrs. Sherry-Anne Hutson
Content Development Assistants: Saira Varghese K. and Ahalya N.

Content Reviewers:
Dr. Rajasekara Murthy C.
Dr. Giridhar D.P.
Dr. Ramachandra T.V.
Dr. Krishna Alluri

Instructional Design Reviewer: Dr. P.R. Ramanujam
Project Coordinator: Dr. T.V. Ramachandra
Project Manager: Dr. Krishna Alluri

ISBN: 978-81-7993-187-5

Published by
TERI Press
The Energy and Resources Institute
Darbari Seth Block
IHC Complex, Lodhi Road
New Delhi – 110 003
India

Tel. 2468 2100 or 4150 4900
Fax 2468 2144 or 2468 2145
India +91 • Delhi (0) 11
E-mail teripress@teri.res.in
Web www.teriin.org

Typeset by TERI Press, New Delhi
Printed in India at Innovative Designers and Printers, New Delhi

Contents

Commonwealth of Learning

The Commonwealth of Learning (COL) is the only international intergovernmental agency that focusses exclusively on using technology to expand the scope and scale of human learning. It operates on the premise that knowledge is the key to individual freedom and to cultural, social and economic development.

COL helps Governments to develop policies that make innovation sustainable and to build systems or applications that expand learning. It works in partnership with other international and bilateral organisations working on the Millenium Development Goals. These include: the Canadian International Development Agency (CIDA), the UK's Department for International Development (DFID), New Zealand's Agency for International Development (NZAID), the Food and Agriculture Organization of the UN (FAO), the World Health Organization (WHO), the United Nations Educational, Scientific and Cultural Organization (UNESCO), the United Nations High Commission for Refugees (UNHCR), The Joint United Nations Programme on HIV/AIDS (UNAIDS), the New Partnership for African Development (NEPAD), the African Virtual University (AVU), the regional Development Banks and the World Bank.

COL is a small agency. It achieves high impact through its focus on technology; placing special emphasis on open and distance learning (ODL) because of its proven effectiveness.

The Chairman of the Board of Governors is HE, the Hon'ble Burchell Whiteman, OJ, and COL's President and Chief Executive Officer is Sir John Daniel.

www.col.org

Indian Institute of Science

http://www.iisc.ernet.in
http://ces.iisc.ernet.in/energy

The Indian Institute of Science (IISc) was started in 1909 through the pioneering vision of J.N. Tata. Since then, it has grown into a premier institution of research and advanced instruction, with more than 2000 active researchers working in almost all frontier areas of science and technology. IISc is an institute of higher learning and is constantly in pursuit of excellence. It is one of the oldest and finest centres of its kind in India, and has a very high international standing in the academic world as well.

The Institute currently has more than forty academic departments, pursuing research in different areas of science and engineering. The departments belong to one of the two faculties of the Institute, *viz.*, the **Faculty of Science** and the **Faculty of Engineering**. Faculty of Science awards PhD degrees through the Regular Research and Integrated PhD programme. On the other hand, Faculty of Engineering awards ME, MTech and MDes degrees.

The Institute has pioneered advanced education in India, and has been making many significant contributions to frontier areas of research. The number of students in the institute is kept deliberately small in order to focus on quality. It has been able to innovate and introduce (a) new systems of imparting knowledge; and (b) educational reforms such as offering courses under unit system.

Preface

Due to rapid increase in the production and consumption processes, societies generate as well as reject solid materials regularly from various sectors — agricultural, commercial, domestic, industrial and institutional. The considerable volume of wastes thus generated and rejected is called solid wastes. In other words, solid wastes are the wastes arising from human and animal activities that are normally solid and are discarded as useless or unwanted. This inevitably places an enormous strain on natural resources and seriously undermines efficient and sustainable development. One of the ways to salvage the situation is through efficient management of solid wastes, and this is the focus of this Course, Management of Municipal Solid Waste. In the 10 units that constitute this Course, we will discuss the processes involved in the management of solid wastes — from waste generation to final disposal.

In Unit 1, we will describe solid wastes and introduce the classification of solid wastes and functional elements, such as waste generation, storage, collection, transport, processing, recovery and disposal, in the management of solid wastes. In Units 2 to 7, we will explain with the support of case studies each of these functional elements. In Unit 8, we will explain the treatment of solid wastes by incineration and energy recovery from the incineration process. Subsequently, in Unit 9, we will deal with the treatment and management of hazardous (biomedical) wastes. Finally, in Unit 10, we will discuss the concept of integrated waste management.

The primary goals of the Course are to encourage reduction of waste at the source and to foster implementation of integrated solid waste management systems that are cost-effective and protect human health and the environment. This Course has been developed for solid waste management practitioners, such as local government officials, teaching professionals, facility owners and operators, consultants, and regulatory agency authorities. This would aid practitioners to meet the daily challenges of planning, managing and operating municipal solid waste management (MSWM) programmes and facilities. The information presented would help decision makers consider the numerous factors associated with successful implementation of solid waste management solutions considering the changes in the infrastructure and technology for handling MSW. The Course would help to carefully evaluate all of the elements in waste handling systems and implement source reduction, recycling, and environmentally sound disposal. When waste generation is unavoidable, the materials can be viewed as a resource from which reusable materials, raw feedstock, minerals, organic matter, nutrients and energy can be recovered for beneficial uses. Residual materials requiring disposal must be carefully managed to protect human health and the environment. The Course would aid all individuals involved with MSWM to expand their professional skills and help other practitioners and community members better understand the challenges we face and the opportunities available to us. It is primarily through the participation of all stakeholders — governments, communities and institutions — that we can make the best possible decisions for the reduction and **integrated management of municipal solid waste**.

T.V. Ramachandra

UNIT 1

Introducing Municipal Solid Waste Management

STRUCTURE

OVERVIEW

In this Unit, we will begin discussion on classification of solid wastes. It touches upon some basics of solid waste management and the need for environmentally sound management practices. It closes with a discussion of the current solid waste management (SWM) scenario in India by presenting the effect of temporal conditions on the quantity of waste generated, analysing the quantity and composition of urban wastes in nine major cities and describing the current disposal methods and their influence on public health and environment.

Before you read any further, note that for the purpose of this Course, we use the terms *solid wastes* and *solid waste* interchangeably.

LEARNING OBJECTIVES

After completing this Unit, you should be able to:

- classify solid wastes;

- explain the functional elements of SWM;
- assess the current situation of SWM in India.

1.1 CLASSIFICATION OF SOLID WASTES

Solid wastes are the organic and inorganic waste materials such as product packaging, grass clippings, furniture, clothing, bottles, kitchen refuse, paper, appliances, paint cans, batteries, etc., produced in a society, which do not generally carry any value to the first user(s). Solid wastes, thus, encompass both a heterogeneous mass of wastes from the urban community as well as a more homogeneous accumulation of agricultural, industrial and mineral wastes. While wastes have little or no value in one setting or to the one who wants to dispose them, the discharged wastes may gain significant value in another setting. Knowledge of the sources and types of solid wastes as well as the information on composition and the rate at which wastes are generated/disposed is, therefore, essential for the design and operation of the functional elements associated with the management of solid wastes.

Solid wastes are classified on the basis of source of generation and type. We will explain these in Subsections 1.1.1 and 1.1.2, respectively.

1.1.1 Source-based Classification

Historically, the sources of solid wastes have been consistent, dependent on sectors and activities (Tchobanoglous et al., 1977), and these include the following:

(i) **Residential:** This refers to wastes from dwellings, apartments, etc., and consists of leftover food, vegetable peels, plastic, clothes, ashes, etc.

(ii) **Commercial:** This refers to wastes consisting of leftover food, glasses, metals, ashes, etc., generated from stores, restaurants, markets, hotels, motels, auto-repair shops, medical facilities, etc.

(iii) **Institutional:** This mainly consists of paper, plastic, glasses, etc., generated from educational, administrative and public buildings such as schools, colleges, offices, prisons, etc.

(iv) **Municipal:** This includes dust, leafy matter, building debris, treatment plant residual sludge, etc., generated from various municipal activities like construction and demolition, street cleaning, landscaping, etc. (Note, however, in India *municipal* can typically subsume items at (i) to (iii) above).

(v) **Industrial:** This mainly consists of process wastes, ashes, demolition and construction wastes, hazardous wastes, etc., due to industrial activities.

(vi) **Agricultural:** This mainly consists of spoiled food grains and vegetables, agricultural remains, litter, etc., generated from fields, orchards, vineyards, farms, etc.

It is important to define the various types of solid wastes that are generated from various sources (see Subsection 1.1.2), which we will do, next.

1.1.2 Type-based Classification

Classification of wastes based on types, i.e., physical, chemical, and biological characteristics of wastes, is as follows (Phelps et al., 1995):

(i) **Garbage:** This refers to animal and vegetable wastes resulting from the handling, sale, storage, preparation, cooking and serving of food. Garbage comprising these wastes contains putrescible (rotting) organic matter, which produces an obnoxious odour and attracts rats and other vermin. It, therefore, requires special attention in storage, handling and disposal.

(ii) **Ashes and residues:** These are substances remaining from the burning of wood, coal, charcoal, coke and other combustible materials for cooking and heating in houses, institutions and small industrial establishments. When produced in large quantities, as in power-generation plants and factories, these are classified as industrial wastes. Ashes consist of fine powdery residue, cinders and clinker often mixed with small pieces of metal and glass. Since ashes and residues are almost entirely inorganic, they are valuable in landfills. (We will discuss landfills in Unit 4.)

(iii) **Combustible and non-combustible wastes:** These consist of wastes generated from households, institutions, commercial activities, etc., excluding food wastes and other highly putrescible material. Typically, while *combustible material* consists of paper, cardboard, textile, rubber, garden trimmings, etc., *non-combustible material* consists of such items as glass, crockery, tin and aluminium cans, ferrous and non-ferrous material and dirt.

(iv) **Bulky wastes:** These include large household appliances such as refrigerators, washing machines, furniture, crates, vehicle parts, tyres, wood, trees and branches. Since these household wastes cannot be accommodated in normal storage containers, they require a special collection mechanism.

(v) **Street wastes:** These refer to wastes that are collected from streets, walkways, alleys, parks and vacant plots, and include paper, cardboard, plastics, dirt, leaves and other vegetable matter. Littering in public places is indeed a widespread and acute problem in many countries including India, and a solid waste management system must address this menace appropriately.

(vi) **Biodegradable and non-biodegradable wastes:** *Biodegradable wastes* mainly refer to substances consisting of organic matter such as leftover food, vegetable and fruit peels, paper, textile, wood, etc., generated from various household and industrial activities. Because of the action of micro-organisms, these wastes are degraded from complex to simpler compounds. *Non-biodegradable* wastes consist of inorganic and recyclable materials such as plastic, glass, cans, metals, etc. Table 1.1 below shows a comparison of biodegradable and non-biodegradable wastes with their degeneration time, i.e., the time required to break from a complex to a simple biological form.

Table 1.1 Biodegradable and Non-Biodegradable Wastes: Degeneration Time

Category	*Type of waste*	*Approximate time taken to degenerate*
Biodegradable	Organic waste such as vegetable and fruit peels, leftover foodstuff, etc.	A week or two
	Paper	10–30 days
	Cotton cloth	2–5 months
	Woollen items	1 year
	Wood	10–15 years
Non-biodegradable	Tin, aluminium, and other metal items such as cans	100–500 years
	Plastic bags	One million years
	Glass bottles	Undetermined

From Table 1.1, we can easily deduce the environmental consequences associated with non-biodegradable wastes such as plastics, glass, etc., which we will discuss later in Unit 6.

(vii) **Dead animals:** With regard to municipal wastes, dead animals are those that die naturally or are accidentally killed on the road. Note that this category does not include carcasses and animal parts from slaughter-houses, which are regarded as industrial wastes. Dead animals are divided into two groups—large and small. Among the large animals are horses, cows, goats, sheep, pigs, etc., and among the small ones are dogs, cats, rabbits, rats, etc. The reason for this differentiation is that large animals require special equipment for lifting and handling when they are removed. If not collected promptly, dead animals pose a threat to public health since they attract flies and other vermin as they decay. Their presence in public places is particularly offensive from the aesthetic point of view as well.

(viii) **Abandoned vehicles:** This category includes automobiles, trucks and trailers that are abandoned on streets and other public places. However, abandoned vehicles have significant scrap value for their metal, and their value to collectors is highly variable.

(ix) **Construction and demolition wastes:** These are wastes generated as a result of construction, refurbishment, repair and demolition of houses, commercial buildings and other structures. They consist mainly of earth, stones, concrete, bricks, lumber, roofing and plumbing materials, heating systems and electrical wires and parts of the general municipal waste stream.

(x) **Farm wastes:** These wastes result from diverse agricultural activities such as planting, harvesting, production of milk, rearing of animals for slaughter and the operation of feedlots. In many areas, the disposal of animal waste has become a critical problem, especially from feedlots, poultry farms and dairies.

(xi) **Hazardous wastes:** *Hazardous wastes* are those defined as wastes of industrial, institutional or consumer origin that are potentially dangerous either immediately or over a period of time to human beings and the environment. This is due to their physical, chemical and biological or radioactive characteristics like ignitability, corrosivity, reactivity and toxicity. Note that in some cases, the active agents may be liquid or gaseous hazardous wastes. These are, nevertheless, classified as solid wastes as they are confined in solid containers. Typical examples of hazardous wastes are empty containers of solvents, paints and pesticides, which are frequently mixed with municipal wastes and become part of the urban waste stream. Certain hazardous wastes may cause explosions in incinerators and fires at landfill sites. Others such as pathological wastes from hospitals and radioactive wastes also require special handling. Effective management practices should ensure that hazardous wastes are stored, collected, transported and disposed of separately, preferably after suitable treatment to render them harmless. We will discuss hazardous wastes in detail in Unit 9.

(xii) **Sewage wastes:** The solid by-products of sewage treatment are classified as sewage wastes. They are mostly organic and derived from the treatment of organic sludge separated from both raw and treated sewages. The inorganic fraction of raw sewage such as grit and eggshells is separated at the preliminary stage of treatment, as it may entrain putrescible organic matter with pathogens and must be buried without delay. The bulk of treated, dewatered sludge is useful as a soil conditioner but is invariably uneconomical. Solid sludge, therefore, enters the stream of municipal wastes, unless special arrangements are made for its disposal.

Table 1.2 summarises our discussion of waste classification based on sources of generation and their types.

Table 1.2 Classification of Solid Wastes

Type	*Description*	*Source*
Garbage	Food waste: wastes from the preparation, cooking and serving of food. Market refuse, waste from the handling, storage, and sale of produce and meat.	Households, institutions and commercial concerns such as hotels, stores, restaurants, markets, etc.
Combustible and non-combustible	Combustible (primary organic) paper, cardboard, cartons, wood, boxes, plastic, rags, cloth, bedding, leather, rubber, grass, leaves, yard trimmings, etc. Non-combustible (primary inorganic) metals, tin, cans, glass bottles, crockery, stones, etc.	
Ashes	Residue from fires used for cooking and for heating building cinders	
Bulky wastes	Large auto parts, tyres, stoves, refrigerators, other large appliances, furniture, large crates, trees, branches, stumps, etc.	
Street wastes	Street sweepings, dirt, leaves, etc.	Streets, sidewalks, alleys, vacant lots, etc.
Dead animals	Dogs, cats, rats, donkeys, etc.	
Abandoned vehicles	Automobiles and spare parts	
Construction and demolition wastes	Roofing, and sheathing scraps, rubble, broken concrete, plaster, conduit pipe, wire, insulation, etc.	Construction and demolition sites.
Industrial wastes	Solid wastes resulting from industry processes and manufacturing operations, such as, food processing wastes, boiler house cinders, wood, plastic and metal scraps, shavings, etc.	Factories, power plants, etc.
Hazardous wastes	Pathological wastes, explosives, radioactive materials, etc.	Households, hospitals, institutions, stores, industry, etc.
Animals and agricultural wastes	Manure, crop residues, etc.	Livestock, farms, feedlots and agriculture
Sewage treatment residue	Coarse screening grit, septic tank sludge, dewatered sludge.	Sewage treatment plants and septic tanks.

Having described solid wastes and their classification, we will now discuss some of the management aspects of solid wastes in Section 1.2. But, before we do so, let us complete Learning Activity 1.1.

LEARNING ACTIVITY 1.1

Based on the source and type, classify the waste generated in your locality.

Note:

a) Write your answer in the space given below.

b) Check your answer with the one given at the end of this book.

1.2 SOLID WASTE MANAGEMENT (SWM)

Solid waste management (SWM) is associated with the control of waste generation, its storage, collection, transfer and transport, processing and disposal in a manner that is in accordance with the best principles of public health, economics, engineering, conservation, aesthetics, public attitude and other environmental considerations.

Put differently, the SWM processes differ depending on factors such as economic status (e.g., the ratio of wealth created by the production of primary products to that derived from manufactured goods, per capita income, etc.), degree of industrialisation, social development (e.g., education, literacy, healthcare, etc.) and quality of life of a location. In addition, regional, seasonal and economic differences influence the SWM processes. This, therefore, warrants management strategies that are economically viable, technically feasible and socially acceptable to carry out such of the functions as are listed below *(http://ces.iisc.ernet.in/energy/SWMTR/TR85.html):*

- Protection of environmental health.
- Promotion of environmental quality.
- Supporting the efficiency and productivity of the economy.
- Generation of employment and income.

SWM has socio-economic and environmental dimensions. In the socio-economic dimension, for example, it includes various phases such as waste storage, collection, transport and disposal, and the management of these phases has to be integrated. In other words, wastes have to be properly stored, collected and disposed off by co-operative management. In addition, poor management of wastes on the user side such as disposing off wastes in the streets, storm water drains, rivers and lakes has to be avoided to preserve the environment, control vector-born diseases and ensure water quality/resource.

Against this background, we will study in Subsection 1.2.1 SWM system.

1.2.1 SWM System

A SWM system refers to a combination of various functional elements associated with the management of solid wastes. The system, when put in place, facilitates the collection and disposal of solid wastes in the community at minimal costs, while preserving public health and ensuring little or minimal adverse impact on the environment. The functional elements that constitute the system are:

(i) **Waste generation:** Wastes are generated at the start of any process, and thereafter, at every stage as raw materials are converted into goods for consumption. The source of waste generation, as we touched upon earlier in Section 1.1, determines quantity, composition and waste characteristics (see Unit 2 for details). For example, wastes are generated from households, commercial areas, industries, institutions, street cleaning and other municipal services. The most important aspect of this part of the SWM system is the identification of waste.

(ii) **Waste storage:** Storage is a key functional element because collection of wastes never takes place at the source or at the time of their generation. The heterogeneous wastes generated in residential areas must be removed within eight days due to shortage of storage space and presence of biodegradable material. Onsite storage is of primary importance due to aesthetic consideration, public health and economics involved. Some of the options for storage are plastic containers, conventional dustbins (of households), used oil drums, large storage bins (for institutions and commercial areas or servicing depots), etc. Obviously, these vary greatly in size, form and material. We shall discuss waste storage in detail in Unit 3.

(iii) **Waste collection:** This includes gathering of wastes and hauling them to the location, where the collection vehicle is emptied, which may be a transfer station (i.e., intermediate station where wastes from smaller vehicles are transferred to larger ones and also segregated), a processing plant or a disposal site. Collection depends on the number of containers, frequency of collection, types of collection services and routes. Typically, collection is provided under various management arrangements, ranging from municipal services to franchised services, and under various forms of contracts.

Note that the solution to the problem of hauling is complicated. For instance, vehicles used for long distance hauling may not be suitable or particularly economic for house-to-house collection. Every SWM system, therefore, requires an individual solution to its waste collection problem, and we will explain this in Unit 3.

(iv) **Transfer and transport:** This functional element involves:

- the transfer of wastes from smaller collection vehicles, where necessary to overcome the problem of narrow access lanes, to larger ones at transfer stations;
- the subsequent transport of the wastes, usually over long distances, to disposal sites.

The factors that contribute to the designing of a transfer station include the type of transfer operation, capacity, equipment, accessories and environmental requirements. We will discuss these in Unit 3.

(v) **Processing:** Processing is required to alter the physical and chemical characteristics of wastes for energy and resource recovery and recycling. The important processing techniques include compaction, thermal volume reduction, manual separation of waste components, incineration and composting. We will discuss the various functions involved in waste processing in detail in Unit 5.

(vi) **Recovery and recycling:** This includes various techniques, equipment and facilities used to improve both the efficiency of disposal system and recovery of usable material and energy. Recovery involves the separation of valuable resources from the mixed solid wastes, delivered at transfer stations or processing plants. It also involves size reduction and density separation by air classifier, magnetic device for iron and screens for glass. The selection of any recovery process is a function of economics, i.e., costs of separation versus the recovered-material products. Certain recovered materials like glass, plastics, paper, etc., can be recycled as they have economic value. We will discuss the various aspects of recovery and recycling, respectively, in Units 6 and 7.

(vii) **Waste disposal:** Disposal is the ultimate fate of all solid wastes, be they residential wastes, semi-solid wastes from municipal and industrial treatment plants, incinerator residues, composts or other substances that have no further use to the society. Thus, land use planning becomes a primary determinant in the selection, design and operation of landfill operations. A modern sanitary landfill is a method of disposing solid waste without creating a nuisance and hazard to public health. Generally, engineering principles are followed to confine the wastes to the smallest possible area, reduce them to the lowest particle volume by compaction at the site and cover them after each day's operation to reduce exposure to vermin. One of the most important functional elements of SWM, therefore, relates to the final use of the reclaimed land.

In Fig. 1.1 below, we show you a typical SWM system with its functional elements and linkages.

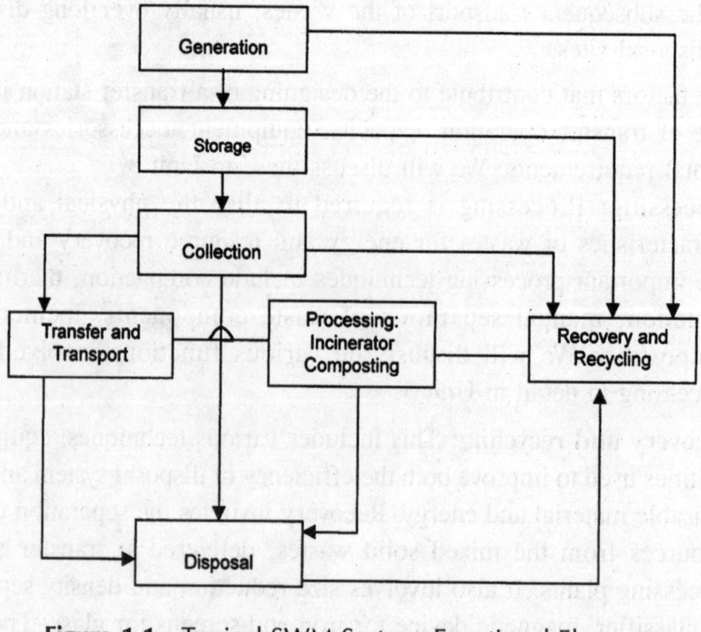

Figure 1.1 Typical SWM System: Functional Elements

You must, however, note that all the elements and linkages shown in Fig. 1.1 need not necessarily be always present in a SWM system. Being generic in its form, this system is applicable to all regions, irrespective of their relative state of development (Tchobanoglous et al., 1977).

> **LEARNING ACTIVITY 1.2**
>
> Which functional elements do you think are needed for SWM in your locality?
>
> **Note:**
>
> a) Write your answer in the space given below.
>
> b) Check your answer with the one given at the end of this book.

1.2.2 ESSWM and EST

We must recognise that each functional element discussed in Subsection 1.2.1 is closely interconnected to minimise adverse impact of wastes on the environment and to maximise the ecosystem carrying capacity. To derive optimal benefits from this, we must apply environmentally sound solid waste management

(ESSWM). This is an integrated approach for controlling and preserving the resources, both in quantity and quality. To improve environmental quality and achieve sustainable development, it is necessary we use EST—environmentally sound technologies (Matsumoto et al., 2000). We will describe ESSWM and EST, next.

Environmentally Sound Solid Waste Management (ESSWM)

In any waste or resource management system, we must pay attention to the interaction between human activities and the ecosystem. We have to recognise that human activities including consumption of goods/services, production of wastes, etc., have a serious impact on the carrying capacity of the ecosystem. This, in turn, affects human health, as the environment deteriorates. The fundamental principles of ESSWM, which take into account economic and social issues along with environmental impact consideration, include the following:

- To ensure sustainable development of the ecosystem and human environment.
- To minimise the impact of human activities on the environment.
- To minimise the impact on the environment and maximise the ecosystem's carrying capacity.
- To ensure the implementation of ESSWM through environmentally sound technologies.

Environmentally Sound Technologies (EST)

EST refers to cost effective and energy efficient technologies, which generally perform better on the environment, as they do not pollute the ecosystem's vital components such as air, land or water and consider the reuse, recycling or recovery of wastes. EST can be categorised broadly as:

- **Hard EST**: This includes equipment, machines and other infrastructure with their material accessories to handle waste products and monitor/measure the quality of air, water and soil.
- **Soft EST**: This supports and complements hard technologies and include *nature-based technologies and management tools*. Nature-based technologies include processes and mechanisms nature uses within a specific ecosystem and its carrying capacity, while management tools include system and procedures, policy and regulatory frameworks, and environmental performance standards and guidelines.

Note that, as implied above, hard and soft technologies complement one another to achieve the goal.

EST is selected based on the following generic criteria, the indicators of which may vary depending on the regions in which they are implemented:

- **Affordability:** This means low investment, reasonableness, maintenance-free and durability.
- **Validity:** This refers to effectiveness, easy operation and maintenance.
- **Sustainability:** This means low impact, energy saving and cultural acceptability.

1.2.3 Factors Affecting SWM System

Many factors influence the decision-making process in the implementation of a SWM system (Phelps et al., 1995). Some of the factors that need to be considered in developing a SWM system are listed below:

(i) **Quantities and characteristics of wastes:** The quantities of wastes generated generally depend on the income level of a family, as higher income category tends to generate larger quantity of wastes, compared to low-income category. The quantity ranges from about 0.25 to about 2.3 kg per person per day, indicating a strong correlation between waste production and per capita income. One of the measures of waste composition (and characteristics) is density, which ranges from 150 kg/m^3 to 600 kg/m^3. Proportion of paper and packaging materials in the waste largely account for the differences. When this proportion is high, the density is low and vice versa. The wastes of high density reflect a relatively high proportion of organic matter and moisture and lower levels of recycling.

(ii) **Climate and seasonal variations:** There are regions in extreme north (> 70° N Latitude) and south (> 60° S Latitude), where temperatures are very low for much of the year. In cold climates, drifting snow and frozen ground interfere with landfill operations, and therefore, trenches must be dug in summer and cover material stockpiled for winter use. Tropical climates, on the other hand, are subject to sharp seasonal variations from wet to dry season, which cause significant changes in the moisture content of solid waste, varying from less than 50% in dry season to greater than 65% in wet months. Collection and disposal of wastes in the wet months are often problematic.

High temperatures and humidity cause solid wastes to decompose far more rapidly than they do in colder climates. The frequency of waste collection in high temperature and humid climates should, therefore, be higher than that in cold climates. In sub-tropical or desert climate, there is no significant variation in moisture content of wastes (due to low rainfall) and low production of leachate from sanitary landfill. High winds and wind blown sand and dust, however, cause special problems at landfill sites. While temperature inversions can cause airborne pollutants to be trapped near ground level, landfill sites can affect groundwater by altering the thermal properties of the soil.

(iii) **Physical characteristics of an urban area:** In urban areas (i.e., towns and cities), where the layout of streets and houses is such that access by vehicles is possible, door-to-door collection of solid wastes is the accepted norm either by large compaction vehicle or smaller vehicle. The picture is, however, quite different in the inner and older city areas where narrow lanes make service by vehicles difficult and often impossible. Added to this is the problem of urban sprawl in the outskirts (of the cities) where population is growing at an alarming rate. Access ways are narrow, unpaved and tortuous, and therefore, not accessible to collection vehicles. Problems of solid waste storage and collection are most acute in such areas.

(iv) **Financial and foreign exchange constraints:** Solid waste management accounts for sizeable proportions of the budgets of municipal corporations. This is allocated for capital resources, which go towards the purchase of equipments, vehicles, fuel and labour costs. Typically, 10% to 40% of the revenues of municipalities are allocated to solid waste management. In regions where wage rates are low, the aim is to optimise vehicle productivity. The unfavourable financial situation of some countries hinders purchase of equipment and vehicles, and this situation is further worsened by the acute shortage of foreign exchange. This means that the balance between the degree of mechanisation and the size of the labour force becomes a critical issue in arriving at the most cost-effective solution.

(v) **Cultural constraints:** In some regions, long-standing traditions preclude the intrusion of waste collection on the precincts of households, and therefore, influence the collection system. In others, where the tradition of caste persists, recruits to the labour force for street cleaning and handling of waste must be drawn from certain sections of the population, while others will not consent to placing storage bins in their immediate vicinity. Social norms of a community more often than not over-ride what many may consider rational solutions. Waste management should, therefore, be sensitive to such local patterns of living and consider these factors in planning, design and operation.

(vi) **Management and technical resources:** Solid waste management, to be successful, requires a wide spectrum of workforce in keeping with the demands of the system. The best system for a region is one which makes full use of indigenous crafts and professional skills and/or ensures that training programmes are in place to provide a self-sustaining supply of trained workforce.

✍ LEARNING ACTIVITY 1.3

Explain the role of environmental sound management and technologies in the context of the SWM system.

Note:

a) Write your answer in the space given below.

b) Check your answer with the one given at the end of this book.

1.3 SWM: THE INDIAN SCENARIO

The problem of municipal solid waste management has acquired alarming dimensions in India especially over the last decade, before which waste management was hardly considered an issue of concern as the waste could be easily disposed off in an environmentally safe manner. However, with time, due to changing lifestyles of people coupled with unplanned developmental activities, urbanisation and industrialisation, the waste quantity and characteristics have changed, and as a result, managing solid wastes has become torturous (Ogawa, 1989; Attarwalla, 1993; Vagale, 1997 and Development, 1999).

The physical and chemical characteristics of Indian city refuse, nonetheless, show that about 80% of it is compostable and ideal for *biogas* generation due to adequate nutrients (NPK), moisture content of 50-55% and a carbon-to-nitrogen ratio of 25-40:1. Therefore, the development of appropriate technologies for utilisation of wastes is essential to minimise adverse health and environmental consequences.

Against this backdrop, let us discuss below the quantum of wastes generated in India, their composition, disposal methods, recycling aspects, and health and environment impacts:

(i) **Waste quantum:** The per capita waste generation rate is about 500 g/day. This along with increased population has contributed to higher total waste generation quantum, as shown in Table 1.3, adding to the problems of municipalities.

Table 1.3 Waste Generation Statistics

Year	*Per capita waste generated (g/day)*	*Total urban municipal waste generated (Mt/year)*
1971	375	14.9
1981	430	25.1
1991	460	43.5
2000	500	48.8

During the last decade, garbage was generated in India at nearly twice the rate of the population growth. Estimates of the solid wastes generated in Indian towns and cities (299 Class I cities with > 100,000 in population and 345 Class II towns) range from 52,000 tonnes to 85,000 tonnes of city garbage every day (while Delhi alone produces 4500-5000 tonnes of municipal solid waste per day). Out of this, only 2832 tonnes get various types of treatment. In addition, studies in nine major metropolitan centres in India indicate that the quantum of wastes generated ranges from 1000 tonnes per day in Patna to 5800 tonnes per day in Mumbai, out of which 80-85% is collected, and the rest is left unattended.

Table 1.4 presents the details about the garbage generated and cleared, and the annual municipal budget in nine major Indian cities.

Table 1.4 Urban Waste Situation in Nine Major Indian Cities

Major Cities	*Garbage Generated (Tonnes per Day)*	*Garbage Cleared (Tonnes per Day)*	*Annual Municipal Budget (Rs. in crores)**
Delhi	3880	2420	1016.28
Kolkata	3500	3150	250.00
Mumbai	5800	5000	2436.00
Bangalore	2130	1800	237.00
Chennai	2675	2140	145.00
Lucknow	1500	1000	48.00
Patna	1000	300	15.00
Ahmedabad	1500	1200	270.00
Surat	1250	1000	170.00

Source: Integrated Modeling of Solid Waste in India (March, 1999) CREED Working Paper Series no 26

* 1 crore = 10 million

(ii) **Waste composition:** Studies reveal that the percentage of the organic matter has remained almost static at 41% in the past three decades, but the recyclables have increased from 9.56% to 17.18% as shown in Table 1.5, below:

Table 1.5 Physico-chemical Characteristics of Municipal Solid Waste

Component	*Percentage on wet weight basis*	
	1971-73 (40 cities)	*1995 (23 cities)*
Paper	4.14	5.78
Plastics	0.69	3.90
Metals	0.50	1.90
Glass	0.40	2.10
Rags	3.83	3.50
Ash and fine earth	49.20	40.30
Total compostable matter	41.24	41.80

Source: Integrated Modeling of Solid Waste in India (March, 1999) CREED Working Paper Series no 26 and CPCB, 1999. Status of Solid Waste Generation, Collection and Disposal in Metropolis (cups/46,1999–2000)

Garbage in Indian cities is estimated to contain about 80% biodegradable waste (as against 25% of US city-garbage) with 50-55% moisture; 35-45% being fruits, vegetable and food biomass; and 8-15% non-organic materials like plastic, metal, glass, stones, etc.

Among various recyclables, plastics have had a quantum jump from 0.69% to 3.9%, i.e., more than a five-fold increase within last twenty years. Plastics due to their unique properties of flexibility, high impact strength, resistance to corrosion and rigidity have replaced valuable natural resources like wood and metals, resulting in a ten-fold increase during the last decade. Of the current consumption of 1.9 million tonnes of plastics, 15% are from the packaging sector, as packaging materials reach the waste bin as a post-consumer waste. Much of this does not have a recycling value, and hence its disposal without any treatment.

Refuse from Indian cities also contains high organic and low combustible matter, if the studies carried out in six cities are of any indication. Presenting the findings of these studies, Table 1.6 shows that the highest organic content is found in Bangalore waste (75%) and the lowest in Kolkata (46%).

Table 1.6 Composition of Urban Solid Waste in Indian Cities (Percentage by weight)

City	*Paper*	*Metals*	*Glass*	*Textiles*	*Plastic**	*Ash and Dust*	*Organics*	*Others***
Chennai	5.90	0.70	-	7.07	-	16.35	56.24	13.74
Delhi	5.88	0.59	0.31	3.56	1.46	22.95	57.71	7.52
Kolkata	0.14	0.66	0.24	0.28	1.54	33.58	46.58	16.98
Bangalore	1.50	0.10	0.20	3.10	0.90	12.00	75.00	7.20
Ahmedabad	5.15	0.80	0.93	4.08	0.69	29.01	48.95	10.39
Mumbai	3.20	0.13	0.52	3.26	-	15.45	59.37	18.07

Source: Integrated Modeling of Solid Waste in India (March, 1999) CREED Working Paper Series No 26. * Includes rubber and leather. ** Includes bones, stones and wooden matter

(iii) **Waste disposal methods:** Waste disposal is the final stage of the waste management cycle. About 90% of the municipal waste collected by the civic authorities in India is dumped in low-lying areas outside the city/town limits, which have no provision of *leachate collection* and treatment, and landfill gas collection and use, as Table 1.7 indicates.

Table 1.7 Waste Disposal Trends in India

Waste disposal method	*1971 (40 cities)*	*1991 (23 cities) and 1999*
Land dumping	Almost all	89.8%
Composting	-	8.6%
Others (Pelletisation, Vermi-composting)	-	1.6%

Source: CPCB, 1999. Status of Solid Waste Generation, Collection and Disposal in Metropolis (cups/46,1999-2000)

As a result, leachate containing heavy metals finds its way to the underground water, rendering it unfit for drinking. The landfill gas escapes into the atmosphere, adding to the greenhouse emissions, which otherwise could be used as thermal fuel. Solid waste management can be an income generating activity with cost benefits (the non-use of materials that would otherwise be required for the two activities, e.g., in the case of power generation, the use of wastes instead of conventional fuels). However, since the Indian city refuse has a low calorific value with high moisture content and quantity of non-combustibles, it is not suitable for incineration, and currently none of the municipal corporations in India runs a full-scale incineration plant. The potential for power generation from urban municipal wastes is tremendous, estimated to generate up to 1000 MW of electricity.

(iv) **Recycling:** This involves collection of recyclables from various sources, which ultimately reach recycling units. It is estimated that about 40-80% of plastic waste gets recycled in India, as compared to 10-15% in the developed nations of the world. However, due to lack of suitable government policies, incentives, subsidies, regulations, standards, etc., related to recycling, this industry is still far behind its western counterparts in terms of technology and quality of manufactured goods. Nevertheless, recycling in India is a highly organised and profit-making venture, though informal in nature.

(v) **Health impacts:** Due to the absence of standards and norms for handling municipal wastes, municipal workers suffer occupational health hazards of waste handling. At the dumpsites in the city of Mumbai, for example, 95 workers were examined and it was found that about 80% of them had eye problems, 73% respiratory ailments, 51% gastrointestinal ailments and 27% skin lesions. Also, municipal workers and rag pickers who operate informally for long hours rummaging through waste also suffer from similar occupational health diseases ranging from respiratory illnesses (from ingesting particulates and bio-aerosols), infections (direct contact with contaminated material), puncture wounds (leading to tetanus, hepatitis and HIV infection) to headaches and nausea, etc. Studies among the 180 rag pickers at open dumps of Kolkata city reveal that average quarterly incidence of diarrhoea was 85%, fever 72% and cough and cold 63%.

(vi) **Environmental impacts:** In addition to occupational health, injury issues and environmental health also need to be mentioned in the context of waste management. Contaminated leachate and surface run-off from land disposal facilities affects ground and surface water quality. Volatile organic compounds and dioxins in air-emissions are attributed to increasing cancer incidence and psychological stress for those living near incinerators or land disposal facilities. Drain clogging due to uncollected wastes leading to stagnant waters and subsequent mosquito vector breeding are a few of the environmental health issues, which affect the waste workers as well as the public. The pneumonic plague that broke out in November 1994 in India (Surat, Gujarat) is a typical example of solid waste mismanagement.

✍ LEARNING ACTIVITY 1.4

What is the composition of waste in your locality and have you seen any significant change during the last few years?

Note:

a) Write your answer in the space given below.

b) Check your answer with the one given at the end of this book.

SUMMARY

In this Unit, we described solid waste and its flow pattern and classified wastes on the basis of source and type, and in that context, discussed residential, commercial, municipal, institutional, industrial and agricultural sectors as major sources of waste generation. Based on the type, we classified wastes as garbage, ashes, street wastes, etc., that indicate the physical, chemical and biological characteristics of wastes. We said that knowledge of these characteristics is important in evaluating alternative equipment needs, systems, and management programmes and plans, especially with respect to the implementation of disposal, and resource and energy recovery options.

We pointed out that the objective of SWM is to minimise the negative environmental effects and identified the functional elements of SWM as waste generation, storage, transfer and transport, processing, reuse and recycling and final disposal. We also said that SWM has economic, social and environmental dimensions. In the context of the socio-economic dimension, we discussed such phases as waste storage, collection, transport and disposal, which are to be integrated. For effective implementation of integrated approaches to management, we said that we require environmentally sound management and technologies for controlling and preserving the resources, both in quantity and quality.

After explaining the functional elements, we discussed the various factors such as quantity and characteristics of waste, climatic and seasonal variations, physical characteristics of urban areas, financial constraints, cultural constraints and management and technical resources, which affect decision-making in solid waste management system. Finally, we illustrated solid waste and its management issues, using the Indian scenario as a case.

SUGGESTED READINGS

Environmental Management in Developing Countries. 1995. *Waste Management, Vol II*. Institute for Scientific Co-operation, Tubingen.

Ali, M., Cotton, A. and Westlake, K. 1999. *Solid Waste Disposal for Low-Income Countries*. Loughborough University, London.

Arne Vesilind, William Worrel and Reinhart Debra. 2002. *Solid Waste Engineering*. Thomson Brooks/Cole, Singapore.

Ramachandra, T.V. and Saira Varghese, K. 2004. Environmentally sound options for e-wastes management. *ENVIS Journal of Human Settlements* (ISSN 0971-9091), March 2004.

Ramachandra, T.V. and Saira Varghese, K. 2003. Exploring possibilities of achieving sustainability in solid waste management. *Indian Journal Environmental Health*, **45(4)**: 255-264.

Tchobanoglous, H., Theisen, H. and Samuel, A. Vigil. 1993. *Integrated Solid Waste Management*. McGraw-Hill, Inc., New Delhi.

US Environmental Protection Agency. 1989. *The Solid Waste Dilemma: An Agenda for Action*. Washington.

US Environmental Protection Agency. 1989. *Decision-Maker's Guide to Solid Waste Management, Vol I*. Washington.

US Environmental Protection Agency. 1995. *Decision-Maker's Guide to Solid Waste Management, Vol II*. Washington.

UNEP. (1996). *International Source Book on Environmentally Sound Technologies for Municipal Solid Waste Management (6)*. IETC, Osaka/Shiga.

http://ces.iisc.ernet.in/energy/SWMTR/TR85.html

http://stratema.sigis.net/cupum/pdf/E1.pdf

REFERENCES

Attarwalla, F.A. 1993. *Solid Waste Management—A Manual*. All India Institute of Local Self-Government, Andheri (W), Bombay, India.

Development, Nagoya, Japan, Vol. 10, No. 3, Autumn, p. xi. Interim Report, 1998. *Solid Waste Management in Class I Cities in India*. Calcutta Municipal Corporation, Calcutta.

Gerlagh, R., Pieter van Beukering, Verma, M., Yadav, P.P. and Pandey, P. March, 1999. *Integrated Modelling of Solid Waste in India*. CREED working paper series no 26.

Matsumoto, S., Santiage, V., Sakaguchi, A. and Yamashiki, Y. 2000. *Environmentally Sound Technologies for Freshwater Management. In:* Proceedings: Waste Recycling and Resource Management in Developing World (Ed: Jana, B.B., Banerjee, R.D., Guterstam, B. and Heeb, J.). University of Kalyani, Calcutta, 617-620.

Ogawa, H. 1989. Selection of Appropriate Technology for Solid Waste Management in Asian Metropolises. *An International Journal of Regional Development Dialogue*. UNCRD, Nagoya, Japan, **10(3)**, Autumn.

Phelps, H.O., Heinke. G. W., Jonker, J.F., Ouano, E.A.R. and Vandecasteele, C. 1995. *Management of Solid Wastes*. UNESCO, Paris.

Pieter van Beukering, Sehker, M., Gerlagh, R. and Kumar, V. March 1999. *Analysing Urban Solid Waste in Developing Countries*. CREED working paper series no 24.

Tchobanoglous, G., Theisen, H. and Eliassan, R. 1977. *Solid Wastes—Engineering Principles and Management Issues*. McGraw-Hill Book Company, New York.

Vagale, L.R. 1997. Environment of Urban Areas in India—Case Study: Bangalore, *ENVIS Journal of Human Settlements*. Centre for Environmental Studies, School of Planning and Architecture, New Delhi, India.

http://ces.iisc.ernet.in/energy/SWMTR/TR85.html

UNIT 2

Waste Generation Aspects

STRUCTURE

OVERVIEW

In Unit 1, we introduced you to the functional elements that constitute a SWM system and suggested that each of these elements would be discussed in detail in Units 2 to 7. We will accordingly discuss waste generation, composition (i.e., components), and characteristics in Unit 2, and will explain the other elements in the subsequent Units. The present Unit thus begins with a discussion on the importance of analysis and field investigations to inform decisions about the planning and design of a SWM system. It then explains waste generation, composition and the physical and chemical characteristics as well as the various factors affecting them. It also outlines the impact of wastes on human health and the environment. It closes with a case study of SWM in Bangalore.

LEARNING OBJECTIVES

After completing this Unit, you should be able to:

- carry out waste stream analysis;
- explain waste generation, composition and the factors affecting them;
- discuss the physical and chemical characteristics of solid wastes;
- determine the effects of poor waste management on public health and the environment.

2.1 WASTE STREAM ASSESSMENT (WSA)

Before we discuss waste generation *per se*, let us first explain the importance of collecting data on waste generation and composition, and the mechanisms that may be used for the purpose.

Waste stream assessment (WSA) is a means to determine the basic aspects of quantity (i.e., the amount of waste generated in the community, both in terms of weight and volume), composition (i.e., the different components of waste stream) and sources of wastes. The information relating to these basic aspects of wastes is vital for making decisions about the SWM system, finance and regulations. Put differently, an assessment of waste stream is essential in the analyses of short- and long-term problems within the local waste management system. It also helps in targeting waste management activities and setting goals for different elements of a waste management plan.

Waste stream assessment, however, is not a one-time activity. It is a continuous and dynamic process, because the characteristics of wastes differ depending on the regions, communities, seasons, etc. We will explain this further in Subsections 2.1.1 and 2.1.2.

2.1.1 Rationale for Analysis

The reasons for the analysis of waste composition, characteristics and quantity include the following (Phelps et al., 1995):

(i) It provides the basic data for the planning, designing and operation of the management systems.

(ii) An ongoing analysis of the data helps detect changes in composition, characteristics and quantities of wastes, and the rates at which these changes take place, which facilitates effective implementation of management systems.

(iii) It quantifies the amount and type of materials suitable for processing, recovery and recycling.

(iv) It provides information that helps in deciding appropriate technologies and equipment.

(v) The forecast trends assist designers and manufacturers in the production of collection vehicles and equipment suitable for future needs.

In the absence of a reliable basic data, carrying out field investigations becomes necessary (Phelps et al., 1995).

2.1.2 Field Investigation

Field investigations may take any one or a combination of the following forms:

(i) **Waste sorting:** Sorting of wastes into predetermined components takes place at disposal sites for weighing and sampling in order to determine the percentage of each component and the physical and chemical characteristics of wastes. It is carried out manually, and the sample size for analysis is between 100 and 150 kg. The implements required for the purpose include sorting table, measuring box, bins or boxes to contain sorted materials and platform weighing machine.

(ii) **Vehicle weighing:** Vehicles are weighed when they enter the disposal sites loaded, and exit the sites empty. The vehicle's front wheels are weighed first, followed by the rear wheels and the sum of the two gives the total weight. Weighing is carried out each day of the weighing period in order to determine the average weight. The weighing of loaded and unloaded vehicles is accomplished with a weighing scale or weighbridge. Ideally, the weighing scale should be operated during the entire period of operation of the disposal site, round the clock, if necessary. An electronic or a mechanical portable axle scale, with a capacity of 20 tonnes, is suitable for the purpose. An electronic scale comprises two load-cell platforms and an electronic control and a display unit. The quantity of waste measured at disposal sites reflects a disposal factor rather than a generation factor, since the measurements do not include wastes that are:

- salvaged at the generation and disposal sites;
- disposed off in unauthorised places such as vacant plots, alleys, ditches, etc.;
- salvaged by collectors;
- lost during transport.

(iii) **Field visits:** This means visiting institutional and industrial sites to identify wastes being generated and disposal methods. Field visits involve visiting the facility, i.e., industry, institutions, etc., viewing the waste handling system and completing a questionnaire with the assistance of the plant manager or senior technical personnel who usually investigate wastes from industries and institutions. Collection of samples in sealed polythene bags follows for laboratory analysis to identify physical and chemical characteristics. Each sample may be in the range of 1.5 to 5 kg.

An assessment of waste stream, in essence, helps us identify components that require improvement for effective implementation of waste management programmes (EPA, 1989 and 1995).

✍ LEARNING ACTIVITY 2.1

State the importance of waste stream assessment (WSA) and how WSA is carried out in your locality.

Note:

a) Write your answer in the space given below.
b) Check your answer with the one given at the end of this book.

Having explained the importance of waste stream assessment, we now discuss how wastes are generated and their composition.

2.2 WASTE GENERATION AND COMPOSITION

Information on waste quantity and composition is important in evaluating alternatives in terms of equipment, systems, plans and management programmes.

For example, if wastes generated at a commercial facility consist of only paper products, the appropriate equipment are shredders and balers. Similarly, on the basis of quantity generated, we can plan appropriate means for separation, collection and recycling programmes. That is to say, the success of SWM depends on the appropriate assessment of quantity of wastes generated. We will elaborate on the aspects of waste generation and composition, respectively, in Subsections 2.2.1 and 2.2.2.

2.2.1 Waste Generation

Waste generation encompasses those activities in which waste, be it solid or semi-solid material, no longer has sufficient economic value for its possessor to retain it.

The processing of raw materials is the first stage when wastes are generated, and waste generation continues thereafter at every step in the process as raw materials are converted into final products for consumption. Figure 2.1 below shows a simplified material-flow diagram indicating the path of generation of solid wastes (Tchobanoglous et al., 1977):

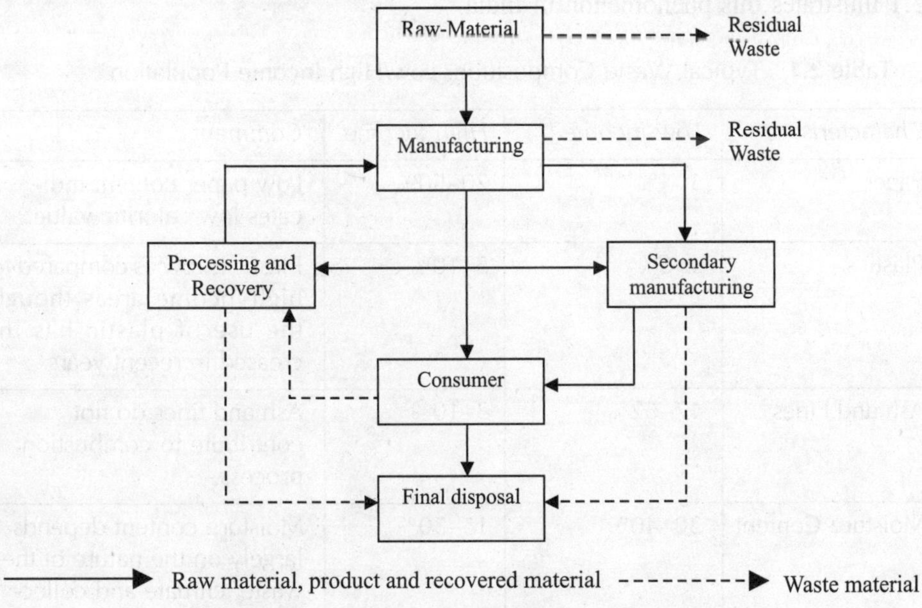

Figure 2.1 Material Flow and Waste Generation

Figure 2.1 suggests that we can reduce the amount of solid waste by limiting the consumption of raw materials and increasing the rate of recovery and reuse. There needs to be, therefore, a societal change in the perception of wastes. This sounds simple. But, implementing changes in the society is difficult, unless appropriate management solutions are provided. That said, we also must note that the changes in waste generation contribute to changes in waste composition, and this we will discuss next.

2.2.2 Waste Composition

Some of the general observations associated with the composition of wastes include the following:

- The major constituents are paper and decomposable organic materials.
- More often than not, metal, glass, ceramics, textile, dirt and wood form part of the composition, and their relative proportion depends on local factors.
- Average proportions of the constituents reaching the disposal sites are consistent and urban wastes are fairly constant although subject to long-term changes such as seasonal variations.

Waste composition varies with the socio-economic status within a particular community, since income, for example, determines life style, composition pattern and cultural behaviour (*http://ces.iisc.ernet.in/energy/SWMTR/TR85.html*). Table 2.1 illustrates this phenomenon in India.

Table 2.1 Typical Waste Composition: Low/High Income Population

Characteristics	*Low income*	*High income*	*Comments*
Paper	1–4%	20–50%	Low paper content indicates low calorific value.
Plastics	1–6%	5–10%	Plastic is low as compared to high-income areas though the use of plastic has increased in recent years.
Ash and Fines	17–62%	3–10%	Ash and fines do not contribute to combustion process.
Moisture Content	30–40%	15–30%	Moisture content depends largely on the nature of the waste, climate and collection frequency. Waste can dry out while awaiting collection.
Bulk Density	300–400 kg/m^3	150 kg/m^3	Heavier waste may cost more to handle and difficult to burn.

Waste composition also depends on the moisture content, density and relative distribution of municipal wastes, as shown in Table 2.2 below, and is important for the characterisation of solid waste for most applications (Ali et al., 1999).

Table 2.2 Solid Wastes: Typical Composition, Moisture and Density

Components	*Mass %*		*Moisture content %*		*Density in kg/m*	
	Range	*Typical*	*Range*	*Typical*	*Range*	*Typical*
Food wastes	6-26	14	50-80	70	120-480	290
Paper	15-45	34	4-10	6	30-130	85
Cardboard	3-15	7	4-8	5	30-80	50
Plastics	2-8	5	1-4	2	30-130	65
Textiles	0-4	2	6-15	10	30-100	65
Rubber	0-1	0.5	1-4	2	90-200	130
Leather	0-2	0.5	8-12	10	90-260	160
Garden trimming	0-20	12	30-80	60	60-225	105
Wood	1-4	2	15-40	20	120-320	240
MIsc. organic substances	0-5	2	10-60	25	90-360	240
Glass	4-16	8	1-4	2	160-480	195
Tin cans	2-8	6	2-4	3	45-160	90
Non-ferrous metals	0-1	1	2-4	2	60-240	160
Ferrous metals	1-4	2	2-6	3	120-1200	320
Dirt, ash, bricks, etc.	0-10	4	6-12	8	320-960	480

Note that the density of waste changes as it moves from the source of generation to the point of ultimate disposal, and such factors as storage methods, salvaging activities, exposure to weather, handling methods and decomposition influence the density. In short, predicting changes of waste composition is as difficult as forecasting waste quantities.

2.2.3 Factors Causing Variation

As we know, wastes cause pollution. While the nature of wastes determines the type and intensity of pollution, it also helps us decide on the appropriate application, engineering design and technology for management. For example, the nature of wastes has implications for collection, transport and recycling. For effective SWM, therefore, we not only need information about the present but also the expected future quantity and composition of wastes.

There are several factors, which affect the present as well as the future waste quantity and composition (Tchobanoglous et al., 1977), and some of which are listed below:

- **Geographic location:** The influence of geographic location is related primarily to different climates that can influence both the amount of certain types of solid wastes generated and the collection operation. For instance, substantial variations in the amount of yard and garden wastes generated in various parts of India are related to the climate. To illustrate, in the warmer southern areas, where the growing season is considerably longer compared to the northern areas, yard wastes are collected in considerably larger quantities and over a longer period of time.
- **Seasons:** Seasons of the year have implications for the quantities and composition of certain types of solid wastes. For example, the growing season of vegetables and fruits affect the quantities of food wastes.
- **Collection frequency:** A general observation is that in localities, where there are ultimate collection services, more wastes are *collected*. Note that this does not mean that more wastes are *generated*. For example, if a homeowner has access to only one or two containers per week, due to limited container capacity, he or she will store newspapers or other materials in some specified storage area. However, the same homeowner will tend to throw them away, if there is access to unlimited container services. In this latter situation, the quantity of waste generated may actually be the same but the quantity collected, as it relates to the frequency of collection, is considerably different.
- **Population diversity:** The characteristics of the population influence the quantity and composition of waste generated. The amount of organic fraction of waste generated is more in low-income areas compared to that in high-income areas. Similarly, the composition differs in terms of paper and other recyclables, which are typically more in high-income areas as against low-income areas (see Table 2.1).
- **Extent of salvaging and recycling:** The existence of salvaging and recycling operation within a community definitely affects the quantity of wastes collected.
- **Public attitude:** Significant reduction in the quantity of solid waste is possible, if and when people are willing to change – on their own volition – their habits and lifestyles to conserve the natural resources and to reduce the economic burden associated with the management of solid wastes.
- **Legislation:** This refers to the existence of local and state regulations concerning the use and disposal of specific materials and is an important factor that influences the composition and generation of certain types of wastes. The Indian legislation dealing with packing and beverage container materials is an example.

✍ LEARNING ACTIVITY 2.2

Discuss the factors that contribute to the variations in composition of solid wastes.

Note:

a) Write your answer in the space given below.

b) Check your answer with the one given at the end of this book.

Having discussed waste generation and its composition in Sections 2.1 and 2.2, respectively, we will introduce in Section 2.3 the physical and chemical characteristics of wastes that generally inform decisions regarding waste management including disposal methods.

2.3 WASTE CHARACTERISTICS

In order to identify the exact characteristics of municipal wastes, it is necessary that we analyse them using physical and chemical parameters (Phelps et al., 1995), which we will discuss in Subsections 2.3.1 and 2.3.2, respectively.

2.3.1 Physical Characteristics

Information and data on the physical characteristics of solid wastes are important for the selection and operation of equipment and for the analysis and design of disposal facilities. The required information and data include the following:

(i) **Density:** Density of waste, i.e., its mass per unit volume (kg/m^3), is a critical factor in the design of a SWM system, e.g., the design of sanitary landfills, storage, types of collection and transport vehicles, etc. To explain, an efficient operation of a landfill demands compaction of wastes to optimum density. Any normal compaction equipment can achieve reduction in volume of wastes by 75%, which increases an initial density of 100 kg/m^3 to 400 kg/m^3. In other words, a waste collection vehicle can haul four times the weight of waste in its compacted state than when it is uncompacted. A high initial density of waste precludes the achievement of a high compaction ratio and the compaction ratio achieved is no greater than 1.5:1. Significant changes in density occur spontaneously as the waste moves from source to disposal, due to scavenging, handling, wetting and drying by the weather, vibration in the collection vehicle and decomposition. Note that:

- the effect of increasing the moisture content of the waste is detrimental in the sense that dry density decreases at higher moisture levels;
- soil-cover plays an important role in containing the waste;
- there is an upper limit to the density, and the conservative estimate of in-place density for waste in a sanitary landfill is about 600 kg/m^3.

(ii) **Moisture content:** Moisture content is defined as the ratio of the weight of water (wet weight – dry weight) to the total weight of the wet waste. Moisture increases the weight of solid wastes, and thereby, the cost of collection and transport. In addition, moisture content is a critical determinant in the economic feasibility of waste treatment by incineration, because wet waste consumes energy for evaporation of water and in raising the temperature of water vapour. In the main, wastes should be insulated from rainfall or other extraneous water. We can calculate the moisture percentage, using the formula given below:

$$\text{Moisture content (\%)} = \frac{\text{Wet weight} - \text{Dry weight}}{\text{Wet wight}} \times 100$$

A typical range of moisture content is 20 to 40%, representing the extremes of wastes in an arid climate and in the wet season of a region of high precipitation. However, values greater than 40% are not uncommon.

(iii) **Size:** Measurement of size distribution of particles in waste stream is important because of its significance in the design of mechanical separators and shredders. Generally, the results of size distribution analysis are expressed in the manner used for soil particle analysis. That is to say, they are expressed as a plot of particle size (mm) against percentage, less than a given value.

LEARNING ACTIVITY 2.3

Determine the moisture content of the sample from the following data:

Waste Composition	% by Weight	Moisture Content
Food wastes	15	70
Paper	45	6
Cardboard	10	5
Plastic	10	2
Garden trimmings	10	60
Wood	5	20
Tin, cans	3	3

Based on 100 kg's

Hint: Compute dry weight (i.e., individual constituent wet weight – moisture). Moisture content of the sample (wet weight – dry weight) × 100/ wet weight.

Note:

a) Write your answer in the space given below.

b) Check your answer with the one given at the end of this book.

2.3.2 Chemical Characteristics

Knowledge of the classification of chemical compounds and their characteristics is essential for the proper understanding of the behaviour of waste, as it moves through the waste management system. The products of decomposition and heating values are two examples of chemical characteristics. If solid wastes are to be used as fuel, or are used for any other purpose, we must know their chemical characteristics, including the following:

(i) **Lipids:** This class of compounds includes fats, oils and grease, and the principal sources of lipids are garbage, cooking oils and fats. Lipids have high heating values, about 38,000 kJ/kg (kilojoules per kilogram), which makes waste with high lipid content suitable for energy recovery. Since lipids become liquid at temperatures slightly above ambient, they add to the liquid content during waste decomposition. Though they are biodegradable, the rate of biodegradation is relatively slow because lipids have a low solubility in water.

(ii) **Carbohydrates:** These are found primarily in food and yard wastes, which encompass sugar and polymer of sugars (e.g., starch, cellulose, etc.) with general formula $(CH_2O)_x$. Carbohydrates are readily biodegraded to products such as carbon dioxide, water and methane. Decomposing carbohydrates attract flies and rats, and therefore, should not be left exposed for long duration.

(iii) **Proteins:** These are compounds containing carbon, hydrogen, oxygen and nitrogen, and consist of an organic acid with a substituted amine group (NH_2). They are mainly found in food and garden wastes. The partial decomposition of these compounds can result in the production of amines that have unpleasant odours.

(iv) **Natural fibres:** These are found in paper products, food and yard wastes and include the natural compounds, cellulose and lignin, that are resistant to biodegradation. (Note that paper is almost 100% cellulose, cotton over 95% and wood products over 40%.) Because they are a highly combustible solid waste, having a high proportion of paper and wood products, they are suitable for incineration. Calorific values of oven-dried paper products are in the range of 12,000 -18,000 kJ/kg and of wood about 20,000 kJ/kg, i.e., about half that for fuel oil, which is 44,200 kJ/kg.

(v) **Synthetic organic material (Plastics):** Accounting for 1–10%, plastics have become a significant component of solid waste in recent years. They are highly resistant to biodegradation and, therefore, are objectionable and of special concern in SWM. Hence the increasing attention being paid to the recycling of plastics to reduce the proportion of this waste component at disposal sites. Plastics have a high heating value, about 32,000 kJ/kg, which makes them very suitable for incineration. But, you must note that polyvinyl chloride (PVC), when burnt, produces dioxin and acid gas. The latter increases corrosion in the combustion system and is responsible for acid rain.

(vi) **Non-combustibles:** This class includes glass, ceramics, metals, dust and ashes, and accounts for 12–25% of dry solids.

(vii) **Heating value:** An evaluation of the potential of waste material for use as fuel for incineration requires a determination of its heating value, expressed as kilojoules per kilogram (kJ/kg). The heating value is determined experimentally using the *Bomb calorimeter test*, in which the heat generated, at a constant temperature of 25 °C from the combustion of a dry sample is measured. Since the test temperature is below the boiling point of water

(100 °C), the combustion water remains in the liquid state. However, during combustion, the temperature of the combustion gases reaches above 100 °C, and the resultant water is in the vapour form. Table 2.3 shows the typical inert residue and heating values for the components of municipal solid waste (Tchobanoglous et al., 1977).

Table 2.3 Typical Heating and Inert Residue Values

Component	*Inert Residue %*		*Heating Value (kJ/kg)*	
	Range	*Typical*	*Range*	*Typical*
Food wastes	2–8	5	3500 – 17000	4500
Paper	4–8	6	11500 –18500	16500
Cardboard	3–6	5	14000 –17500	16000
Plastics	2–20	10	28000 –37000	32500
Textiles	2–4	2.5	15000 –20000	17500
Rubber	8–20	10	21000 –28000	18500
Leather	8–20	10	15000 –20000	17500
Garden trimmings	2–6	4.5	2300 –18500	6500
Wood	0.6–2	1.5	17500 –20000	18500
Glass	96–99	98	120 –240	140
Tin cans	96–99	96	—	—
Nonferrous metals	90–99	96	240 –1200	700
Ferrous metals	94–99	98	240 –1200	700
Dirt, ash, bricks, etc.	60–80	70	2300 –11500	7000
Municipal solid waste			9500 –13000	10500

Note that while evaluating incineration as a means of disposal or energy recovery, we need to consider the heating values of respective constituents (see Table 2.3). For example:

- Organic material yields energy only when dry.
- The moisture content in the waste reduces the dry organic material per kilogram of waste and requires a significant amount of energy for drying.
- The ash content of the waste reduces the proportion of dry organic material per kilogram of waste and retains some heat when removed from the furnace.

(viii) **Ultimate analysis:** This refers to an analysis of waste to determine the proportion of carbon, hydrogen, oxygen, nitrogen and sulphur, and the analysis is done to make mass balance calculation for a chemical or thermal process. Besides, it is necessary to determine ash fraction because of its potentially harmful environmental effects, brought about by the presence of toxic metals such as cadmium, chromium, mercury, nickel, lead, tin and zinc. Note that other metals (e.g., iron, magnesium, etc.) may also be present but they are non-toxic. Table 2.4 shows the result of ultimate analysis of a typical municipal solid waste.

Table 2.4 Municipal Solid Waste: A Typical Ultimate Analysis

Element	*Range (%dry weight)*
Carbon	25-30
Hydrogen	2.5-6.0
Oxygen	15-30
Nitrogen	0.25-1.2
Sulphur	0.02-0.12
Ash	12-30

(ix) **Proximate analysis:** This is important in evaluating the combustion properties of wastes or a waste or refuse derived fuel. The fractions of interest are:

- moisture content, which adds weight to the waste without increasing its heating value, and the evaporation of water reduces the heat released from the fuel;
- ash, which adds weight without generating any heat during combustion;
- volatile matter, i.e., that portion of the waste that is converted to gases before and during combustion; and
- fixed carbon, which represents the carbon remaining on the surface grates as charcoal. A waste or fuel with a high proportion of fixed carbon requires a longer retention time on the furnace grates to achieve complete combustion than a waste or fuel with a low proportion of fixed carbon.

Table 2.5 illustrates a proximate analysis for the combustible components of municipal solid waste.

Table 2.5 Municipal Solid Waste: A Typical Proximate Analysis

Components	*Value, percent*	
	Range	*Typical*
Moisture	15-40	20
Volatile matter	40-60	53
Fixed carbon	5-12	7
Glass, metal, ash	15-30	20

What Subsection 2.3.2 implies is that to evaluate alternative processing and recovery options (e.g., incineration process), we need information on the chemical characteristics of wastes, and wastes can typically be a combination of combustible and non-combustible materials.

✍ LEARNING ACTIVITY 2.4

Identify and list the chemical characteristics of two types of municipal wastes commonly present in your locality.

Note:

a) Write your answer in the space given below.

b) Check your answer with the one given at the end of this book.

Waste mismanagement, e.g., the practice of throwing wastes into streets, storm water drains, vacant land, etc., leads to breeding of disease vectors such as rats, with their attendant fleas carrying germs, etc., which results in an epidemic such as plague, malaria, etc. This has an adverse impact on public health and environment. We will look into the consequences of waste mismanagement on public health and the environment in Section 2.4.

2.4 HEALTH AND ENVIRONMENTAL EFFECTS

An effective solid waste management system is necessary to avoid public health disasters, spread of disease by insects and vectors and adverse effect on water and air (Phelps et al., 1995). Solid waste workers are the most exposed to the risks of parasitic infections and accidents, and therefore, a SWM system must include proper mechanisms to avoid these incidences. To the direct and indirect risks through accidents, exposure and spread of disease, we must add the effect of visual pollution caused by litter and nuisance created by smoke and dust at disposal sites. Against this background, let us discuss the health and environmental effects of solid wastes in Subsections 2.4.1 and 2.4.2.

2.4.1 Public Health Effect

The volume of waste is increasing rapidly as a result of increasing population and improving economic conditions in various localities. This increased volume of wastes is posing serious problems due to insufficient workforce and other constraints in disposing it off properly. What are the consequences of improper management and handling of wastes? Consider the following:

(i) **Disease vectors and pathways:** Wastes dumped indiscriminately provide the food and environment for thriving populations of vermin, which are the agents of various diseases. The pathways of pathogen transmission from wastes to humans are mostly indirect through insects – flies, mosquitoes and roaches and animals – rodents and pigs. Diseases become a public health problem when they are present in the human and animal population of surrounding communities, or if a carrier transmits the etiological agent from host to receptor.

(ii) **Flies:** Most common in this category is the housefly, which transmits typhoid, salmonellosis, gastro-enteritis and dysentery. Flies have a flight range of about 10 km, and therefore, they are able to spread their influence over a relatively wide area. The four stages in their life-cycle are egg, larva, pupa and adult. Eggs are deposited in the warm, moist environment of decomposing food wastes. When they hatch, the larvae feed on the organic material, until certain maturity is reached, at which time they migrate from the waste to the soil of other dry loose material before being transformed into pupae. The pupae are inactive until the adult-fly emerges. The migration of larvae within 4 to 10 days provides the clue to an effective control measure, necessitating the removal of waste before migration of larvae. Consequently, in warm weather, municipal waste should be collected twice weekly for effective control. In addition, the quality of household and commercial storage containers is very significant. The guiding principle here is to restrict access to flies. Clearly, the use of suitable storage containers and general cleanliness at their location, as well as frequent collection of wastes, greatly reduces the population of flies. Control is also necessary at transfer stations, composting facilities and disposal sites to prevent them from becoming breeding grounds for flies. Covering solid wastes with a layer of earth at landfill sites at the end of every day arrests the problem of fly breeding at the final stage.

(iii) **Mosquitoes:** They transmit diseases such as malaria, filaria and dengue fever. Since they breed in stagnant water, control measures should centre on the elimination of breeding places such as tins, cans, tyres, etc. Proper sanitary practices and general cleanliness in the community help eliminate the mosquito problems caused by the mismanagement of solid waste.

(iv) **Roaches:** These cause infection by physical contact and can transmit typhoid, cholera and amoebiasis. The problems of roaches are associated with the poor storage of solid waste.

(v) **Rodents:** Rodents (rats) proliferate in uncontrolled deposits of solid wastes, which provide a source of food as well as shelter. They are responsible for the spread of diseases such as plague, murine typhus, leptospirosis, histoplasmosis, rat bite fever, dalmonelosis, trichinosis, etc. The fleas, which rats carry, also cause many diseases. This problem is associated not only with open dumping but also poor sanitation.

(vi) **Occupational hazards:** Workers handling wastes are at risk of accidents related to the nature of material and lack of safety precautions. The sharp edges of glass and metal and poorly constructed storage containers may inflict injuries to workers. It is, therefore, necessary for waste handlers to wear gloves, masks and be vaccinated. The infections associated with waste handling include:

- skin and blood infections resulting from direct contact with waste and from infected wounds;
- eye and respiratory infections resulting from exposure to infected dust, especially during landfill operations;
- diseases that result from the bites of animals feeding on the waste;
- intestinal infections that are transmitted by flies feeding on the waste;
- chronic respiratory diseases, including cancers resulting from exposure to dust and hazardous compounds.

In addition, the accidents associated with waste handling include:

- bone and muscle disorders resulting from the handling of heavy containers and the loading heights of vehicles;

- infecting wounds resulting from contact with sharp objects;
- reduced visibility, due to dust along the access routes, creates greater risk of accidents;
- poisoning and chemical burns resulting from contact with small amounts of hazardous chemical wastes mixed with general wastes such as pesticides, cleaning solutions and solvents in households and commercial establishments;
- burns and other injuries resulting from occupational accidents at waste disposal sites or from methane gas explosion at landfill sites;
- serious health hazards, particularly for children, due to careless dumping of lead-acid, nickel-cadmium and mercuric oxide batteries.

(vii) **Animals:** Apart from rodents, some animals (e.g., dogs, cats, pigs, etc.) also act as carriers of disease. For example, pigs are involved in the spread of diseases like trichinosis, cysticerosis and toxoplasmosis, which are transmitted through infected pork, eaten either in raw state or improperly cooked. Solid wastes, when fed to pigs, should be properly treated (cooked at 100 °C for at least 50 minutes with suitable equipment).

2.4.2 Environmental Effect

Besides causing health disorders that we have touched upon in Subsection 2.4.1, inadequate and improper waste management causes adverse environmental effects such as the following:

(i) **Air pollution:** Burning of solid wastes in open dumps or in improperly designed incinerators emit pollutants (gaseous and particulate matters) to the atmosphere. Studies show that the environmental consequences of open burning are greater than incinerators, especially with respect to aldehydes and particulates. Emissions from an uncontrolled incinerator system include particulate matter, sulphur oxides, nitrogen oxides, hydrogen chloride, carbon monoxide, lead and mercury. Discharge of arsenic, cadmium and selenium is to be controlled, since they are toxic at relatively low exposure levels. Polychlorinated dibenzofurans (PCDFs), commonly called dioxins and furans, are of concern because of their toxicity, carcinogenicity and possible mutagenicity.

(ii) **Water and land pollution:** Water pollution results from dumping in open areas and storm water drains, and improper design, construction and/or operation of a sanitary landfill. Control of infiltration from rainfall and surface runoff is essential in order to minimise the production of leachate. Pollution of groundwater can occur as a result of:

- the flow of groundwater through deposits of solid waste at landfill sites;

- percolation of rainfall or irrigation waters from solid wastes to the water table;
- diffusion and collection of gases generated by the decomposition of solid wastes.

The interaction between leachate contaminants and the soil depends on the characteristics of the soil. Soil bacteria stabilise biological oxygen demand (BOD), i.e., the amount of oxygen required by micro-organisms to degrade organic matter, by anaerobic action, if toxic substances are in low concentration. The carbon dioxide produced keeps the pH level low, causing the water to dissolve minerals in the aquifers. Consequently, the change in groundwater quality may take place depending on the characteristics of the aquifer. Contamination can spread over considerable distances from the landfill, if the aquifers are of sand or gravel. In clayey soils, the rate of movement is greatly reduced. The capacity of clay to exchange ions restricts the movement of metal ions by capturing them in the soil matrix. Changes in its chemical characteristics are due to hardness, iron and manganese compounds.

(iii) **Visual pollution:** The aesthetic sensibility is offended by the unsightliness of piles of wastes on the roadside. The situation is made worse by the presence of scavengers rummaging in the waste. Waste carelessly and irresponsibly discarded in public thoroughfares, along roads and highways and around communal bins (i.e., makeshift containers, without lids, used for the storage of residential, commercial and institutional wastes) gives easy access to animals scavenging for food. The solution to this social problem undoubtedly lies in the implementation of public education at all levels – primary, secondary, tertiary and adult, both short- and long-term, and in raising the status of public health workers and managers in solid waste management.

(iv) **Noise pollution:** Undesirable noise is a nuisance associated with operations at landfills, incinerators, transfer stations and sites used for recycling. This is due to the movement of vehicles, the operation of large machines and the diverse operations at an incinerator site. The impacts of noise pollution may be reduced by careful siting of SWM operations and by the use of noise barriers.

(v) **Odour pollution:** Obnoxious odours due to the presence of decaying organic matter are characteristic of open dumps. They arise from anaerobic decomposition processes and their major constituents are particularly offensive. Proper landfill covering eliminates this nuisance.

(vi) **Explosion hazards:** Landfill gas, which is released during anaerobic decomposition processes, contains a high proportion of methane (35–73%). It can migrate through the soil over a considerable distance, leaving the buildings in the vicinity of sanitary landfill sites at risk, even after the closure of landfills. Several methods are available for control of landfill gas, such as venting, flaring and the use of impermeable barriers.

✍ LEARNING ACTIVITY 2.5

List the adverse health and environment impacts due to improper handling of solid waste. Identify at least four such effects in your locality.

Note:

a) Write your answer in the space given below.
b) Check your answer with the one given at the end of this book.

2.5 CASE STUDY: STATUS OF WASTE GENERATION IN BANGALORE

SWM in urban areas has interface with all aspects of life and government administration (Areivala, 1971). And, through this case study, we intend to share with you some practical aspects of SWM, from the disposal of building debris to organic, putrescible, and bio-medical wastes and their possible recycling potential. For details, see *http://stratema.sigis.net/cupum/pdf/E1.pdf*.

Bangalore, also known as the Garden City, is one of the fastest growing metropolitan cities in South India. It is the state capital of Karnataka and the sixth largest city in India. Topographically, Bangalore is located in the south deccan and physically, has grown on watershed running through the middle of the Mysore Plateau from west to east which serves as the main water parting of the state at an average elevation of 900 metres above sea level. The city gets moderate rainfall of around 900 mm largely between June and October. On account of its elevation, Bangalore is bestowed with salubrious and equable climate comparable to those of temperate regions.

The city covers the local planning area of 500 sq. kms. Out of this, 226.16 sq. kms are developed at present. In 1991, its population was 4.16 million and has grown over six million, as per recent projected population estimates. Besides, the city has a floating population of over 0.5 million. It is a fast growing city beset with the usual problems of inadequate waste management, due to constraints such as lack of finance and other resources, deficiencies in equipment and workforce and paucity of space (for waste disposal).

The Bangalore Mahanagara Palike (BMP), the erstwhile Bangalore City Corporation (BCC), is concerned with the prime areas of public health, solid waste management including health care waste and sanitation, education, horticulture, etc. Working with non-governmental and voluntary organisations of all stakeholders in a participatory approach, BMP is striving to implement suggestions towards an improved SWM.

SWM is a vital function of the Health Department, supported by the Engineering Department of BMP and the health officer heads the SWM wing. In the present context, SWM in BMP essentially means the cleaning of streets, emptying dustbins, transportation of wastes to city outskirts and burning them in open areas for their disposal (Attarwalla, 1993; Gotoh, 1989; Development, 1998; Ogawa, 1989; Vagale, 1997).

The waste generation and composition details of Bangalore are as follows:

(i) **Waste generation:** Bangalore produces over 2500 tonnes of solid waste per day and the Municipal Corporation has miserably inadequate infrastructure in managing the disposal of solid wastes generated. It is estimated that the per capita generation of solid waste works out to 0.59 kg/day. The sources of waste generation and the amount generated at each source are given in Table 2.6.

Table 2.6 Different Sources of Solid Waste Generation in Bangalore

Sl No.	*Source*	*Quantity (in MT/day)*
1	Households	1000
2	Shops, Establishments, Institutions, etc.	600
3	Markets	600
4	Others	300
5	TOTAL	2500

Source: Department of SWM, Bangalore Mahanagara Palike, Bangalore

(ii) **Waste composition:** The composition of wastes in Bangalore has wide variations in the proportion of contents. It varies from area to area, depending upon the socio-economic conditions and the population density. The composition of the total wastes generated in Bangalore city is given in Table 2.7.

Table 2.7 Composition of Solid Waste in Bangalore

Sl No.	*Type of Waste*	*Composition (in percentage)*
1	Putrescible waste	75.2
2	Dust and ash	12.0
3	Textiles	3.1
4	Paper	1.5
5	Plastic, leather and rubber	0.9
6	Glass	0.2
7	Metals	0.1
8	Earth and building debris and others	0.7

Source: Department of SWM, Bangalore Mahanagara Palike, Bangalore

In Bangalore, there are 401 slum pockets identified which come under the jurisdiction of three different authorities, viz., Bangalore City Corporation - 64 slums, Bangalore Development Authority – 64 slums, and Karnataka Slum Clearance Board – 273 slums. Fifteen percent of the city's population lives in these slums (Comprehensive Development Plan (Revised) Bangalore Report, 1995, p. 25). The slum locations are generally found to be least desirable from the point of view of habitation – being low lying areas, tank beds, quarry pits, near railway lines and cemeteries. The authorities have been unable to clear the garbage from most of the slums mainly due to the slum dwellers' practice of throwing their wastes into drains, and only part of the waste generated is available for collection. There are 12 large vegetable and fruit markets other than a number of small groups of pavement vegetable vendors. Approximately, these markets are producing more than 150 tonnes of wastes daily. In addition to this, large quantities of wastes get generated from slaughterhouses, food packing industries and cold storage facilities.

SUMMARY

This Unit began with a discussion on the significance of collecting information and data on wastes, and in that context discussed some aspects of waste stream assessment that would help in the planning and design of waste management activities. Following this discussion, we explained waste generation and composition and pointed out that solid wastes generated vary in composition and characteristics depending on the source of generation (domestic, industries, agriculture, institution or commercial sectors) and factors such as geographic location, seasonal variation, collection frequency, public attitude, etc. Then, we discussed how the information on the physical and chemical characteristics of wastes would help decide the various elements of waste management and disposal options. Subsequently, we discussed the consequences of improper disposal of solid waste on public health and the environment (e.g., air, water, visual, noise and odour pollution, explosion hazards, etc.). We then touched upon some of these aspects through a case study of Bangalore.

SUGGESTED READINGS

Tchobanoglous, G., Theisen, H. and Eliassan, R. 1977. *Solid Wastes—Engineering Principles and Management Issues*. McGraw-Hill Book Company, New York.

Arne Vesilind, William Worrel and Reinhart Debra. 2002. *Solid Waste Engineering*, Thomson Brooks/Cole, Singapore.

Tchobanoglous, G., Theisen, H. and Samuel, A. Vigil. 1993. *Integrated Solid Waste Management*. McGraw-Hill, Inc., New Delhi.

Neissen, W. 1977. Properties of waste materials. *In:* D.G. Wilson, ed., *Handbook of Solid Waste Management*. Van Nostrand Reinhold, New York.

American Society of Testing and Materials. 1992. *A Standard Test Method for Determination of the Composition of Unprocessed Municipal Solid Waste*. D 5231-92, Philadelphia.

UNEP. 1996. *International Source Book on Environmentally Sound Technologies for Municipal Solid Waste Management (6),* IETC, Osaka/Shiga.

http://ces.iisc.ernet.in/energy/SWMTR/TR85.html

http://stratema.sigis.net/cupum/pdf/E1.pdf

REFERENCES

Ali, S.M., Cotton, A.P. and Westlake, K. 1999. *Down to Earth: Solid Waste Disposal for Low-Income Countries*. WEDC, Loughborough University, UK.

Areivala, S.J. 1971. *Solid Wastes Disposal in India*. Central/Public Health Engineering Institute, Nagpur.

Attarwalla, F.A. 1993. *Solid Waste Management - A Manual*. All India Institute of Local Self-Government, Andheri (W), Bombay, India.

Development, Nagoya, Japan, Vol. 10, No. 3, autumn, p. xi. Interim Report. 1998. *Solid Waste Management in Class I Cities in India*. Calcutta Municipal Corporation, Calcutta.

Gotoh, S. 1989a. Issues and Factors to be Considered for Improvement of Solid Waste Management in Asian Metropolises. *Regional Development Dialogue,* **10(3):** 1-10.

Gotoh, S. 1989b. Solid Waste Management: A base for healthier Metropolitan Development, Editorial Introduction, *Regional Development Dialogue*. **10(3):** i-ii.

Ogawa, H. 1989. *Selection of Appropriate Technology for Solid Waste Management in Asian Metropolises*. An International Journal of Regional Development Dialogue, UNCRD, Nagoya, Japan, Vol. 10, No. 3, Autumn.

Phelps, H.O., Heinke. G.W., Jonker, J.F., Ouano, E.A.R. and Vandecasteele, C. 1995. *Management of Solid Wastes*. UNESCO, Paris.

Tchobanoglous, G., Theisen, H. and Eliassan, R. 1977. *Solid Wastes—Engineering Principles and Management Issues*. McGraw-Hill Book Company, New York.

US Environmental Protection Agency. 1989. *Decision-Maker's Guide to Solid Waste Management, Vol I*. Washington.

US Environmental Protection Agency. 1995. *Decision-Maker's Guide to Solid Waste Management, Vol II*. Washington.

Vagale, L.R. 1997. *Environment of Urban Areas in India—Case Study: Bangalore*, ENVIS Journal of Human Settlements. Centre for Environmental Studies, School of Planning and Architecture, New Delhi, India, November 1997.

http://ces.iisc.ernet.in/energy/SWMTR/TR85.html

UNIT 3

Waste Collection, Storage and Transport

STRUCTURE

OVERVIEW

For the final disposal of the wastes generated (see Unit 2), it is imperative that we put in place an effective waste collection system, which we described in Unit 1 (see (iii) of Subsection 1.2.1). In Unit 3, we will build on this description and discuss in detail the various aspects of collection system. Accordingly, we will first explain the components of waste collection such as storage, collection crew, route, transfer station, etc. We will then discuss each of these components. We will also discuss the design, operation and implementation of waste collection systems. We will close the Unit with a case study highlighting waste storage, collection and transport.

LEARNING OBJECTIVES

After completing this Unit, you should be able to:

- discuss the various components of a waste collection system;
- explain the characteristics of waste containers relative to their use;
- state the purpose of a transfer station;
- evaluate how a collection system is planned and implemented;
- collect and maintain the required data for record keeping and inventory control;
- design and implement a collection system.

3.1 COLLECTION COMPONENTS

As described in Subsection 1.2.1, Unit 1, waste collection does not mean merely the gathering of wastes, and the process includes, as well, the transporting of wastes to transfer stations and/or disposal sites. To elaborate, the factors that influence the waste collection system include the following (EPA, 1989 and Ali, et al., 1999):

(i) **Collection points:** These affect such collection system components as crew size and storage, which ultimately control the cost of collection. Note that the collection points depend on locality and may be residential, commercial or industrial.

(ii) **Collection frequency:** Climatic conditions and requirements of a locality as well as containers and costs determine the collection frequency. In hot and humid climates, for example, solid wastes must be collected at least twice a week, as the decomposing solid wastes produce bad odour and leachate. And, as residential wastes usually contain food wastes and other putrescible (rotting) material, frequent collection is desirable for health and aesthetic reasons. Besides climates, the quality of solid waste containers on site also determines the collection frequency. For instance, while sealed or closed containers allow collection frequency up to three days, open and unsealed containers may require daily collection. Collection efficiency largely depends on the demography of the area (such as income groups, community, etc.), where collection takes place. While deciding collection frequency, therefore, you must consider the following:

- cost, e.g., optimal collection frequency reduces the cost as it involves fewer trucks, employees and reduction in total route distance;
- storage space, e.g., less frequent collection may require more storage space in the locality;

- sanitation, e.g., frequent collection reduces concerns about health, safety and nuisance associated with stored refuse.

(iii) **Storage containers** (see also Subsection 3.2.1)**:** Proper container selection can save collection energy, increase the speed of collection and reduce crew size. Most importantly, containers should be functional for the amount and type of materials and collection vehicles used. Containers should also be durable, easy to handle, and economical, as well as resistant to corrosion, weather and animals. In residential areas, where refuse is collected manually, standardised metal or plastic containers are typically required for waste storage. When mechanised collection systems are used, containers are specifically designed to fit the truck-mounted loading mechanisms. While evaluating residential waste containers, consider the following:

- efficiency, i.e., the containers should help maximise the overall collection efficiency.
- convenience, i.e., the containers must be easily manageable both for residents and collection crew.
- compatibility, i.e., the containers must be compatible with collection equipment.
- public health and safety, i.e., the containers should be securely covered and stored.
- ownership, i.e., the municipal ownership must guarantee compatibility with collection equipment.

(iv) **Collection crew** (see also Subsection 3.3.1)**:** The optimum crew size for a community depends on labour and equipment costs, collection methods and route characteristics. The size of the collection crew also depends on the size and type of collection vehicle used, space between the houses, waste generation rate and collection frequency. For example, increase in waste generation rate and quantity of wastes collected per stop due to less frequent collection result in a bigger crew size. Note also that the collection vehicle could be a motorised vehicle, a pushcart or a trailer towed by a suitable prime mover (tractor, etc.). It is possible to adjust the ratio of collectors to collection vehicles such that the crew idle time is minimised. However, it is not easy to implement this measure, as it may result in an overlap in the crew collection and truck idle time. An effective collection crew size and proper workforce management can influence the productivity of the collection system. The crew size, in essence, can have a great effect on overall collection costs. However, with increase in collection costs, the trend in recent years is towards:

- decrease in the frequency of collection;

- increase in the dependence on residents to sort waste materials;
- increase in the degree of automation used in collection.

This trend has, in fact, contributed to smaller crews in municipalities.

(v) **Collection route** (see also Subsection 3.3.2)**:** The collection programme must consider the route that is efficient for collection. An efficient routing of collection vehicles helps decrease costs by reducing the labour expended for collection. Proper planning of collection route also helps conserve energy and minimise working hours and vehicle fuel consumption. It is necessary, therefore, to develop detailed route configurations and collection schedules for the selected collection system. The size of each route, however, depends on the amount of waste collected per stop, distance between stops, loading time and traffic conditions. Barriers, such as railroad, embankments, rivers and roads with heavy traffic, can be considered to divide route territories. Routing (network) analyses and planning can:

- increase the likelihood of all streets being serviced equally and consistently;
- help supervisors locate or track crews quickly;
- provide optimal routes that can be tested against driver judgement and experience.

(vi) **Transfer station** (see also Section 3.4)**:** A transfer station is an intermediate station between final disposal option and collection point in order to increase the efficiency of the system, as collection vehicles and crew remain closer to routes. If the disposal site is far from the collection area, it is justifiable to have a transfer station, where smaller collection vehicles transfer their loads to larger vehicles, which then haul the waste long distances. In some instances, the transfer station serves as a pre-processing point, where wastes are dewatered, scooped or compressed. A centralised sorting and recovery of recyclable materials are also carried out at transfer stations (EPA, 1989). The unit cost of hauling solid wastes from a collection area to a transfer station and then to a disposal site decreases, as the size of the collection vehicle increases. This is due to various reasons such as the following:

- Labour costs remain constant.
- The ratio of payload to vehicle load increases with vehicle size.
- The waiting time, unloading time, idle time at traffic lights and driver rest period are constant, regardless of the collection vehicle size.

✎ LEARNING ACTIVITY 3.1

Find the current waste collection practice in your locality and state its role in waste management.

Note:

a) Write your answer in the space given below.

b) Check your answer with the one given at the end of this book.

Note that waste collection often proves to be the most costly component of any waste management system. However, with a proper collection system design and management, we can significantly reduce the costs. Consider the following criteria to evaluate, and make decisions about, collection systems:

- **Efficiency:** Do the services help minimise the cost per household?
- **Effectiveness:** Do the services satisfy the community needs?
- **Equity:** Do the services address equally the concerns of all social and demographic groups?
- **Reliability:** Do the services ensure consistency?
- **Safety and environmental impact:** Do the services ensure safety of workers, public health and protection of the environment?

Note also that various management arrangements, ranging from municipal services to franchised services, and under various forms of contracts are, typically, in vogue for waste collection. One of the critical decisions to be made at the planning stage, therefore, is as to who – the public or private agencies – operates the collection system, though the final decision depends on the existing conditions and options for the local decision-makers (EPA, 1989).

3.2 STORAGE: CONTAINERS/COLLECTION VEHICLES

As mentioned in Unit 1, waste storage is an important component of a waste management system. Waste storage encompasses proper containers to store wastes and efficient transport of wastes without any spillage to transfer stations/disposal sites. We will analyse these two aspects of waste storage in Subsections 3.2.1 and 3.2.2.

3.2.1 Containers/Storage Bins

The design of an efficient waste collection system requires careful consideration of the type, size and location of containers at the point of generation for storage of wastes until they are collected. While single-family households generally use small containers, residential units, commercial units, institutions and industries require large containers. Smaller containers are usually handled manually whereas the larger, heavier ones require mechanical handling. The containers may fall under either of the following two categories:

(i) Stationary containers: These are used for contents to be transferred to collection vehicles at the site of storage.

(ii) Hauled containers: These are used for contents to be directly transferred to a processing plant, transfer station or disposal site for emptying before being returned to the storage site.

The desirable characteristics of a well-designed container are low cost, size, weight, shape, resistance to corrosion, water tightness, strength and durability (Phelps et al., 1995). For example, a container for manual handling by one person should not weigh more than 20 kg, lest it may lead to occupational health hazards such as muscular strain, etc. Containers that weigh more than 20 kg, when full, require two or more crew members to manually load and unload the wastes, and which result in low collection efficiency.

Containers should not have rough or sharp edges, and preferably have a handle and a wheel to facilitate mobility. They should be covered to prevent rainwater from entering (which increases the weight and rate of decomposition of organic materials) into the solid wastes. The container body must be strong enough to resist and discourage stray animals and scavengers from ripping it as well as withstand rough handling by the collection crew and mechanical loading equipment. Containers should be provided with a lifting bar, compatible with the hoisting mechanism of the vehicle. The material used should be light, recyclable, easily moulded and the surface must be smooth and resistant to corrosion. On the one hand, steel and ferrous containers are heavy and subject to corrosion; the rust peels off exposing sharp edges, which could be hazardous to the collection crew. On the other, wooden containers (e.g., bamboo, rattan and wooden baskets) readily absorb and retain moisture and their surfaces are generally rough, irregular and difficult to clean.

Communal Containers

Generally, the containers used for waste storage are communal containers. Figure 3.1 below shows a typical communal container, which a compactor collection vehicle (see Fig. 3.5) can lift and empty mechanically.

Figure 3.1 Typical Communal Container

The use of communal containers is largely dependent on local culture, tradition and attitudes towards waste. Communal containers may be fixed on the ground (stationary) or movable (hauled). Movable containers are provided with hoists and tails compatible with lifting mechanism of collection vehicles and such containers have capacities of 1–4 m^3. The waste management authority must monitor, maintain and upgrade the communal containers. Note that in residential and commercial areas in India, the communal containers are often made of concrete.

In areas with very high waste generation rates, i.e., rates exceeding two truckloads daily, such as wet markets, large commercial centres and large business establishments, roll-on-roll or hoisted communal containers with capacities of 12–20 m^3 and a strong superstructure with wheels are used. Normally, the collection vehicle keeps an empty container as a replacement before it hauls the filled container. When a truck is used as a collection vehicle, the use of communal containers may be appropriate.

It is advisable to place the containers 100–200 m apart for economic reasons. The communal containers are usually staggered such that the effective distance of 100 m is maintained as shown in Fig. 3.2.

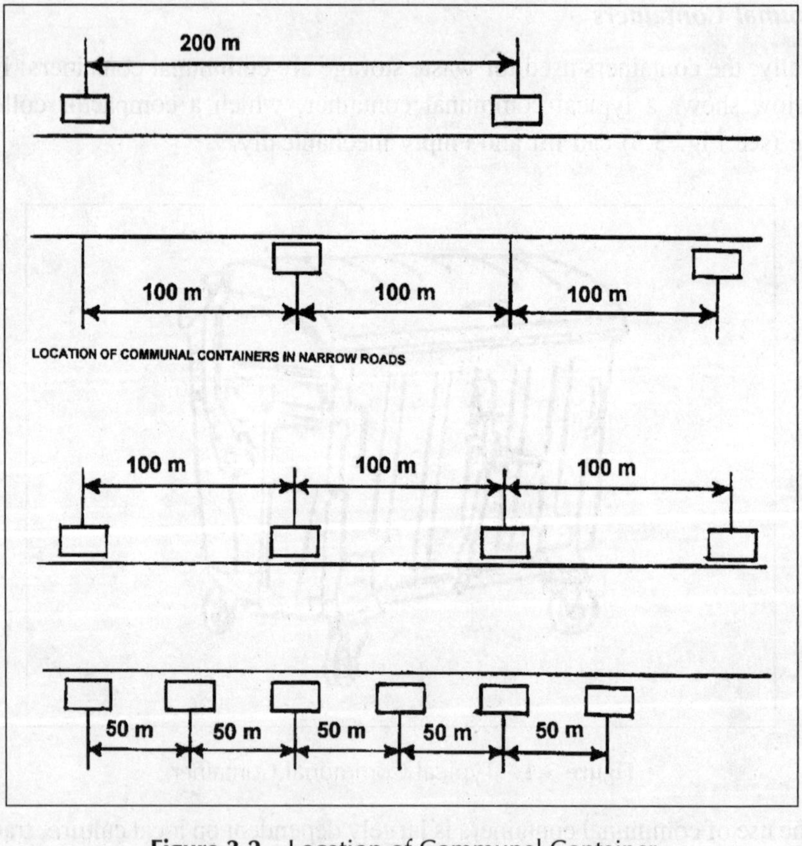

Figure 3.2 Location of Communal Container

This means that the farthest distance the householder will have to walk is 50 metres. However, in narrow streets with low traffic, where the house owner can readily cross the street, a longer distance is advisable. If the collection vehicle has to stop frequently, say, at every 50 m or so, fuel consumption increases, and this must be avoided.

Disadvantages

The major disadvantage of communal containers is the potential lack of maintenance and upgrading. The residuals and scattered solid wastes emit foul odours, which discourage residents from using the containers properly. In addition, if fixed containers are built below the vehicle level, the collection crew may be held responsible for sweeping and loading the solid wastes into transfer containers before being loaded into the collection vehicle. Sweeping and cleaning the communal containers of residuals obviously impinge on the time of the crew members and take a longer time than if the wastes are placed in smaller containers. As fixed communal containers have higher rates of failure, their use is not advisable.

To overcome the problem of maintaining communal containers, individual residents should maintain their own containers and locate them in designated areas. The communal area must have water and drains to facilitate the cleaning of the containers. This practice has the advantage of reducing the number of collection stops and at the same time maintaining the householder's responsibility for cleaning them. The residents must also be properly educated on the importance of good housekeeping as the containers in the communal area are subject to vandalism. In short, if communal containers are to be successful, the design of the containers, loading and unloading areas, and collection vehicle accessories should be co-ordinated.

3.2.2 Collection Vehicles

Almost all collections are based on collector and collection crew, which move through the collection service area with a vehicle for collecting the waste material. The collection vehicle selected must be appropriate to the terrain, type and density of waste generation points, the way it travels and type and kind of material (UNEP, 1996). It also depends upon strength, stature and capability of the crew that will work with it. The collection vehicle may be small and simple (e.g., two-wheeled cart pulled by an individual) or large, complex and energy intensive (e.g., rear loading compactor truck). The most commonly used collection vehicle is the dump truck fitted with a hydraulic lifting mechanism. A description of some vehicle types follows:

(i) **Small-scale collection and muscle-powered vehicles:** These are common vehicles used for waste collection in many countries and are generally used in rural hilly areas. As Fig. 3.3 illustrates, these can be small rickshaws, carts or wagons pulled by people or animals, and are less expensive, easier to build and maintain compared to other vehicles:

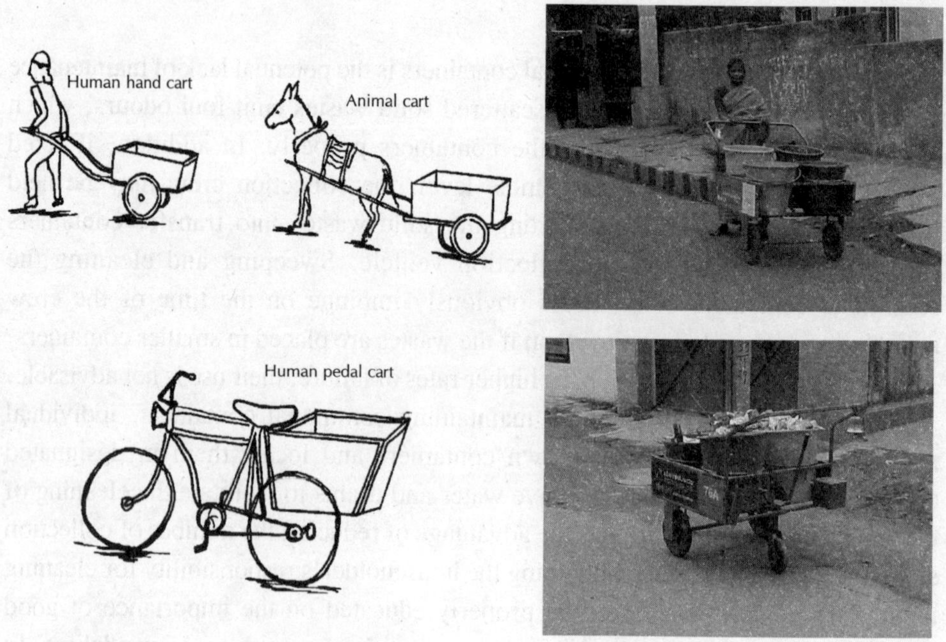

Figure 3.3 Small-scale Collection Vehicles: An Illustration

They are suitable for densely populated areas with narrow lanes, and squatter settlements, where there is relatively low volume of waste generated. Some drawbacks of these collection vehicles include limited travel range of the vehicles and weather exposure that affect humans and animals.

(ii) **Non-compactor trucks:** Non-compactor trucks are efficient and cost effective in small cities and in areas where wastes tend to be very dense and have little potential for compaction. Figure 3.4 illustrates non-compactor trucks.

Figure 3.4 Non-compactor Trucks

When these trucks are used for waste collection, they need a dumping system to easily discharge the waste. It is generally required to cover the trucks in order to prevent residue flying off or rain soaking the wastes. Trucks with capacities of 10–12 m^3 are effective, if the distance between the disposal site and the collection area is less than 15 km. If the distance is longer, a potential transfer station closer than 10 km from the collection area is required. Non-compactor trucks are generally used, when labour cost is high. Controlling and operating cost is a deciding factor, when collection routes are long and relatively sparsely populated.

(iii) **Compactor truck:** Compaction vehicles are more common these days, generally having capacities of 12–15 m^3 due to limitations imposed by narrow roads. Although the capacity of a compaction vehicle, illustrated in Fig. 3.5, is similar to that of a dump truck, the weight of solid wastes collected per trip is 2 to 2.5 times larger since the wastes are hydraulically compacted.

Figure 3.5 Compactor Truck

A compactor truck allows waste containers to be emptied into the vehicle from the rear, front or sides and inhibits vectors (of disease) from reaching the waste during collection and transport. It works poorly when waste stream is very dense, wet, collected materials are gritty or abrasive, or when the roads are dusty. The advantages of the compactor collection vehicle include the following:

- containers are uniform, large, covered and relatively visually inoffensive;
- waste is set out in containers so that the crew can pick them up quickly;
- health risk to the collectors and odour on the streets are minimised;
- waste is relatively inaccessible to the waste pickers.

> ✍ **LEARNING ACTIVITY 3.2**
>
> What are the types of containers and collection vehicles in your locality?
>
> **Note:**
>
> a) Write your answer in the space given below.
>
> b) Check your answer with the one given at the end of this book.

3.3 COLLECTION OPERATION

In Section 3.2, we introduced you to different types of containers and collection vehicles. We now discuss the movement of collection crew in terms of workforce efficiency and collection routes.

3.3.1 Movement of Collection Crew

In cultures such as India, Bangladesh, etc., solid waste collection is assigned to the lowest social group. More often, the collection crew member accepts the job as a temporary position or stopgap arrangement, while looking for other jobs that are considered more respectable.

Apart from this cultural problem, the attitude of some SWM authorities affects collection operation. For example, some authorities still think that the collection of solid waste is mechanical, and therefore, the collection crew does not need any training to acquire special skills. As a result, when a new waste collector starts working, he or she is sent to the field without firm instruction concerning his or her duties, responsibilities and required skills. For an effective collection operation, the collection team must properly be trained. The collection crew and the driver of the collection vehicle must, for example, work as a team, and this is important to maintain the team morale and a sense of social responsibility among these workers.

You must also note that the movement of collection crew, container location and vehicle stopping point affect collection system costs. Figure 3.6 highlights the distance the collection crew will have to walk, if it were to serve the farthest point first or serve the point closest to the vehicle.

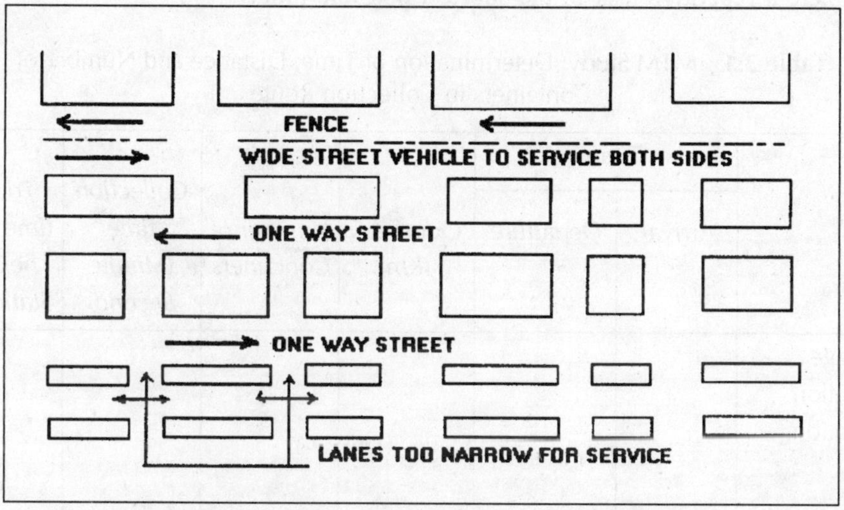

Figure 3.6 Effect of Container Location and Vehicle Stopping

The difference may be one or two minutes per collection stop, but it matters with the number of stops the crew will take in a working shift. Multiplying the minutes by the total number of crew working and labour cost depicts the amount of labour hours lost in terms of monetary value.

Generally, familiarity of the crew with the collection area improves efficiency. For example, the driver becomes familiar with the traffic jams, potholes and other obstructions that he or she must avoid. The crew is aware of the location of the containers and the vehicle stops. It is, therefore, important to assign each crew specific areas of responsibility. Working together also establishes an understanding of the strong and weak points of the team members and efficient work sequences. The collection operation must also observe a strict time schedule. Testing of new routes, new gadgets and vehicles is best carried out first in the laboratory and later in a pilot area. Testing of a new sequence using the whole service area could result in disorder and breakdown of the solid waste collection system. Studies show that it takes two hours to recover for every hour of a failed system.

Motion Time Measurement (MTM) Technique

Motion time measurement (MTM) studies are now an integral part of the standard procedure in the development of solid waste collection systems. MTM is a technique to observe and estimate the movement of the collection crew with the help of stopwatches. The results thus gathered are tabulated as shown in Table 3.1 to determine the best sequence of activities that workers must follow in order to complete a repetitive task in the shortest possible time.

Table 3.1 MTM Study: Determination of Time, Distance and Number of Containers in Collection Route

	Time					
	Arrival	*Departure*	*Odometer (Km)*	*Number of Containers*	*Collection time (Minute Second)*	*Trip time to next Station*
Garage						
1 Station	..	..				
2 Station	..	..				
.	..	..				
.	..	..				
.	..	..				
20 Station	..	..				
Last Station	..	..				
Disposal Site	..	..				
Total	..	..				
Weight	With Load tonne		With Load tonne		With Load tonne	

Source: Phelps et al., 1995

MTM also helps in deciding the best combination of equipment to maintain a desired level of output, reduce health problems related to the repetitive work sequence and predict the effects of changes in materials handled.

Sophisticated MTM studies involve hidden or open video cameras at different collection stops to record, replay and study the operation sequence of the collection crew. If the crew is conscious of being observed, they tend to work faster and reduce time wastage in unauthorised salvaging and other non-scheduled activities. Once the crew is familiar with the person(s) observing them, it begins to perform more credibly. In studies involving video cameras, therefore, the first two or three hours of observation are often neglected.

3.3.2 Collection Vehicle Routing

Efficient routing and re-routing of solid waste collection vehicles can help decrease costs by reducing the labour expended for collection. Routing procedures usually consist of the following two separate components:

(i) **Macro-routing:** Macro-routing, also referred to as route-balancing, consists of dividing the total collection area into routes, sized in such a way as to represent a day's collection for each crew. The size of each route depends on the amount of waste collected per stop, distance between stops, loading time and traffic conditions. Barriers, such as railroad embankments, rivers and roads with heavy competing traffic, can be used to divide route territories. As much as possible, the size and shape of route areas should be balanced within the limits imposed by such barriers.

(ii) **Micro-routing:** Using the results of the macro-routing analysis, micro-routing can define the specific path that each crew and collection vehicle will take each collection day. Results of micro-routing analyses can then be used to readjust macro-routing decisions. Micro-routing analyses should also include input and review from experienced collection drivers.

The heuristic (i.e., trial and error) route development process is a relatively simple manual approach that applies specific routing patterns to block configurations. The map should show collection, service garage locations, disposal or transfer sites, one-way streets, natural barriers and areas of heavy traffic flow. Routes should then be traced onto the tracing paper using the following rules:

- Routes should not be fragmented or overlapping. Each route should be compact, consisting of street segments clustered in the same geographical area.
- Total collection plus hauling time should be reasonably constant for each route in the community.
- The collection route should be started as close to the garage or motor pool as possible, taking into account heavily travelled and one-way streets.
- Heavily travelled streets should not be visited during rush hours.
- In the case of one-way streets, it is best to start the route near the upper end of the street, working down it through the looping process.

- Services on dead-end streets can be considered as services on the street segment that they intersect, since they can only be collected by passing down that street segment. To keep right turns at a minimum (in countries where driving is left-oriented), collection from the dead-end streets is done when they are to the left of the truck. They must be collected by walking down, reversing the vehicle or taking a U-turn.
- Waste on a steep hill should be collected, when practical, on both sides of the street while vehicle is moving downhill. This facilitates safe, easy and fast collection. It also lessens wear of vehicle and conserves gas and oil.
- Higher elevations should be at the start of the route.
- For collection from one side of the street at a time, it is generally best to route with many anti-clockwise turns around blocks.
- For collection from both sides of the street at the same time, it is generally best to route with long, straight paths across the grid before looping anti-clockwise.
- For certain block configurations within the route, specific routing patterns should be applied. (Adapted from American Public Works Association, 1975.)

Based on the above rules, Fig. 3.7 below illustrates a typical collection vehicle routing.

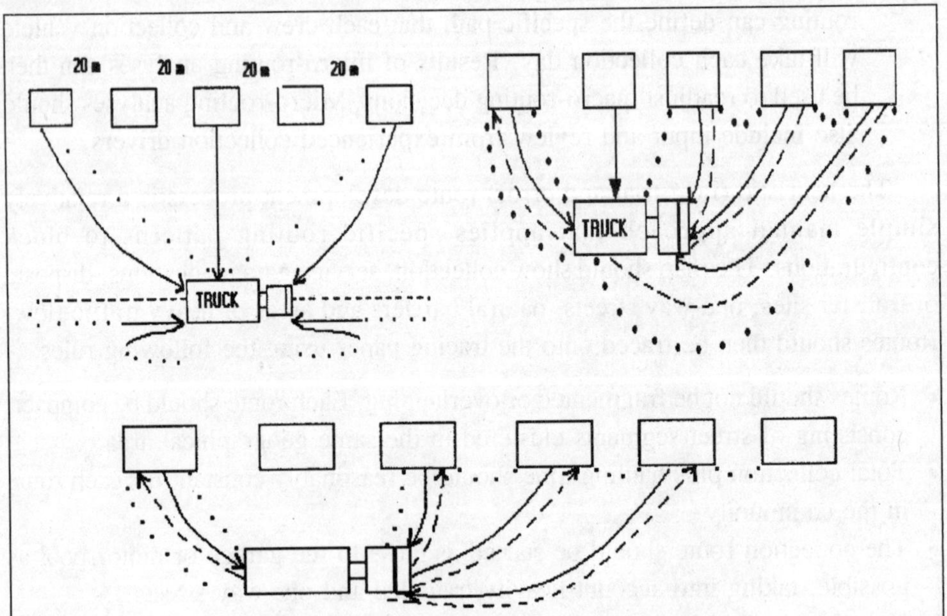

Figure 3.7 Collection Vehicle Route

Before we proceed any further, let us complete Learning Activity 3.3.

✍ LEARNING ACTIVITY 3.3

State the rules that we need to keep in mind, while designing a collection route.

Note:

a) Write your answer in the space given below.

b) Check your answer with the one given at the end of this book.

3.4 TRANSFER STATION

As mentioned earlier in Section 3.1, transfer station is a centralised facility, where waste is unloaded from smaller collection vehicles and re-loaded into large vehicles for transport to a disposal or processing site. This transfer of waste is frequently accompanied by removal, separation or handling of waste. In areas, where wastes are not already dense, they may be compacted at a transfer station. The technical limitations of smaller collection vehicles and the low hauling cost of solid waste, using larger vehicles, make a transfer station viable. Also, the use of transfer station proves reasonable, when there is a need for vehicles servicing a collection route to travel shorter distances, unload and return quickly to their primary task of collecting the waste.

Limitations in hauling solid wastes are the main factors to be considered, while evaluating the use of transfer stations. These include the additional capital costs of purchasing trailers, building transfer stations and the extra time, labour and energy required for transferring wastes from collection truck to transfer trailer.

Consider also the following factors that affect the selection of a transfer station:

- Types of waste received.
- Processes required in recovering material from wastes.
- Required capacity and amount of waste storage desired.
- Types of collection vehicles using the facility.
- Types of transfer vehicles that can be accommodated at the disposal facilities.
- Site topography and access.

The main problem in the establishment of a transfer station, however, is securing a suitable site. Stored solid wastes and recyclable materials, if not properly handled, will attract flies and other insect vectors. Odours from the transferred solid wastes will also be a nuisance, if not properly controlled. In addition, the traffic and noise due to small and large collection vehicles, collectors, drivers, etc., invite the resentment of the communities living in the vicinity of transfer stations (EPA, 1995).

3.4.1 Types

Depending on the size, transfer stations can be either of the following two types:

(i) **Small to medium transfer stations:** These are direct-discharge stations that provide no intermediate waste storage area. The capacities are generally small (less than 100 tonnes/day) and medium (100 to 500 tonnes/day). Depending on weather, site aesthetics and environmental concerns, transfer operations of this size may be located either indoor or outdoor. More complex small transfer stations are usually attended during hours of operation and may include some simple waste and materials processing facilities. For example, it includes a recyclable material separation and processing centre. The required overall station capacity (i.e., the number and size of containers) depends on the size and population density of the area served and the frequency of collection.

(ii) **Large transfer stations:** These are designed for heavy commercial use by private and municipal collection vehicles. The typical operational procedure for a larger station is as follows:

- When collection vehicles arrive at the site, they are checked in for billing, weighed and directed to the appropriate dumping area.

- Collection vehicles travel to the dumping area and empty the wastes into a waiting trailer, a pit or a platform.
- After unloading, the collection vehicle leaves the site, and there is no need to weigh the departing vehicle, if its weight (empty) is known.
- Transfer vehicles are weighed either during or after loading. If weighed during loading, trailers can be more consistently loaded to just under maximum legal weights and this maximises payloads and minimises weight violations.

Designs for Larger Transfer Operations

Several different designs for larger transfer operations are common, depending on the transfer distance and vehicle type. Most designs, however, fall into one of the following three categories:

(i) **Direct-discharge non-compaction station:** In these stations, waste is dumped directly from collection vehicle into waiting transfer trailers and is generally designed with two main operating floors. In the transfer operation, wastes are dumped directly from collection vehicles (on the top floor) through a hopper and into open top trailers on the lower floor. The trailers are often positioned on scales so that dumping can be stopped when the maximum payload is reached. A stationary crane with a bucket is often used to distribute the waste in the trailer. After loading, a cover or tarpaulin is placed over the trailer top. However, some provision for waste storage during peak time or system interruptions should be developed. Because of the use of little hydraulic equipment, a shutdown is unlikely and this station minimises handling of waste.

(ii) **Platform/pit non-compaction station:** In this arrangement, the collection vehicles dump their wastes onto a platform or into a pit using waste handling equipment, where wastes can be temporarily stored, and if desired, picked through for recyclables or unacceptable materials. The waste is then pushed into open-top trailers, usually by front-end loaders. Like direct discharge stations, platform stations have two levels. If a pit is used, however, the station has three levels. A major advantage of these stations is that they provide temporary storage, which allows peak inflow of wastes to be levelled out over a longer period. Construction costs for this type of facility are usually higher because of the increased floor space. This station provides convenient and efficient storage area and due to simplicity of operation and equipment, the potential for station shutdown is less.

(iii) **Compaction station:** In this type of station, the mechanical equipment is used to increase the density of wastes before they are transferred. The most common type of compaction station uses a hydraulically powered compactor to compress wastes. Wastes are fed into the compactor through a chute, either directly from collection trucks or after intermediate use of a pit. The hydraulic ram of the compactor pushes waste into the transfer trailer, which is usually mechanically linked to the compactor (EPA, 1995). Compaction stations are used when:

- wastes must be baled for shipment;
- open-top trailers cannot be used because of size restrictions;
- site topography or layout does not accommodate a multi-level building.

The main disadvantage of a compaction facility is that the facility's ability to process wastes is directly dependent on the operativeness of the compactor. Selection of a quality compactor, regular maintenance of the equipment, easy availability of spare parts and prompt availability of the service personnel are essential for the station's reliable operation.

3.4.2 Capacity

A transfer station should have enough capacity to manage and handle the wastes at the facility throughout its operating life. While selecting the design capacity of a transfer station, we must, therefore, consider trade-offs between the capital costs associated with the station and equipment and the operational costs. Designers should also plan adequate space for waste storage and, if necessary, waste processing. Transfer stations are usually designed to have 1.5–2 days of storage capacity. The collection vehicle unloading area is usually the waste storage area and sometimes a waste sorting area. When planning the unloading area, designers should allow adequate space for vehicle and equipment manoeuvring. To minimise the space required, the facility should be designed such that the collection vehicle backs into the unloading position. Adequate space should also be available for offices, employee facilities, and other facility-related activities (EPA, 1995). Factors that should be considered in determining the appropriate capacity of a transfer facility include:

- capacity of collection vehicles using the facility;
- desired number of days of storage space on tipping floor;
- time required to unload collection vehicles;
- number of vehicles that will use the station and their expected days and hours of arrival;
- waste sorting or processing to be accomplished at the facility;
- transfer trailer capacity;
- hours of station operation;

- availability of transfer trailers waiting for loading;
- time required, if necessary, to attach and disconnect trailers from tractors or compactors.

Transfer station capacity can be determined using the following formulae:

(i) **Pit stations:** Based on the rate at which wastes can be unloaded from collection vehicles:

$$C = P_c \times (L/W) \times (60 \times H_w/T_c) \times F$$

Based on rate at which transfer trailers are loaded:

$$C = (P_t \times N \times 60 \times H_t)/(T_t + B)$$

(ii) **Direct dump stations:**

$$C = (N_n \times P_t \times F \times 60 \times H_w)/[((P_t/P_c) \times (W/L_n)) \times T_c + B]$$

(iii) **Hopper compaction stations:**

$$C = (N_n \times P_t \times F \times 60 \times H_w)/[(P_t/P_c \times T_c) + B]$$

(iv) **Push pit compaction station:**

$$C = (N_p \times P_t \times F \times 60 \times H_w)/[(P_t/P_c \times W/L_p \times T_c) + B_c + B]$$

where C = Station capacity (tonnes/day); P_c = Collection vehicle payload (tonnes); L = Total length of dumping space (feet); H_w = Hours per day that waste is delivered; T_c = Time (in minutes) to unload each collection vehicle; F = Peaking factor (ratio of the number of collection vehicles received during an average 30-minute period to the number rceeived during a peak 30-minute period); L_p = Length of push pit (feet); N_p = Number of push pits; B_c = Total cycle time for clearing each push pit and compacting waste into trailer; P_t = Transfer trailer payload (tonnes); N = Number of transfer trailers loading simultaneously; H_t = Hours per day used to load trailers (minutes); B = Time to remove and replace each loaded trailer (minutes); T_t = Time to load each transfer trailer (minutes); N_n = Number of hoppers; and L_n = Length of each hopper (feet).

These formulae are useful in estimating the capacity of various types of transfer stations (EPA, 1995) and should be adapted, as necessary, for specific applications.

3.4.3 Viability

Transfer stations offer benefits such as lower collection costs (because crews waste less time travelling to the site), reduced fuel and maintenance costs for collection vehicles, increased flexibility in selection of disposal facilities, opportunity to recover recyclables or compostables at the transfer site and the opportunity to shred or scoop wastes prior to disposal. These benefits must be weighed against the costs to develop and operate the facility.

The classical approach to arrive at the economic viability of operating a transfer station, is to add the unit cost of the transfer station to the cost of hauling using large vehicles, and to compare this cost with the cost of hauling directly to the disposal site using the smaller vehicles that service the collection area. The cost of hauling using small vehicles is the sum of the depreciation cost of the vehicle, driver's salary, salary of the collection crew (if they are on standby waiting for the vehicle to return to the collection area) and fuel cost. The transfer station cost is the sum of the transfer station's depreciation cost and the operating and maintenance costs divided by the capacity of the station. The cost of using the large vehicle is the sum of the vehicle depreciation, fuel cost and driver's salary.

The cost-effectiveness of a transfer station depends on the distance of disposal site from the generation area, and a distance of 10–15 km is usually the minimum cost-effective distance (Phelps et al., 1995). The distance between the disposal site and collection area is one of the principal variables in deciding whether to use a transfer station or haul the solid wastes directly from the collection area to the disposal site. Figure 3.8 illustrates the economic analysis involving the effect of the hauling distance on the collection cost.

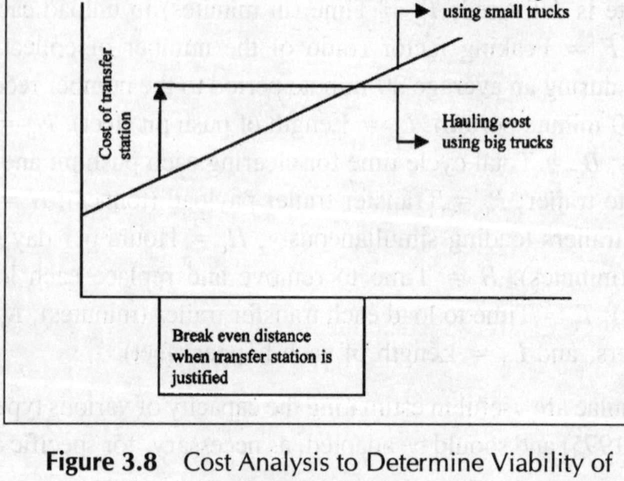

Figure 3.8 Cost Analysis to Determine Viability of Transfer Station

Now, let us consider first the case in which the transfer station is located directly along the hauling route between the disposal site and the collection area. Let the unit cost of hauling using a small vehicle be Rs. A/m^3 km. The cost of operation, maintenance, depreciation, loading and unloading at the transfer station be Rs. B/m^3 and the cost of hauling using large vehicles be Rs. C/m^3 km. If the distance between the collection area and the transfer station is X km and the distance between the transfer station and the disposal site is Y km, then the distance between the collection area and the disposal site is $X + Y$ km. Then, the total cost of hauling the solid wastes from the collection area to the disposal site using a transfer station is:

$$T = 2AX + B + 2CY$$

The factor 2 is added to account for the round trip, which effectively doubles the distance travelled. The total cost of hauling without the transfer station is:

$$T_1 = 2A(X + Y)$$

The transfer station is justified, when:

$$T < T_1$$

That is, the hauling cost using a transfer station is lower than the direct hauling costs between the collection area and the disposal site. Substituting the values of T and T_1 yields:

$$2AX + B + 2CY < 2AX + 2AY$$

or

$$Y > B/(2A - 2C)$$

Note that X cancels out. The distance between the potential transfer station site and the disposal site is the variable to consider. The distance between the collection area and the disposal site is important in deciding the utilisation of a transfer station, if X is equal to zero, in which case the transfer station is located right at the centroid of the collection area. Under normal conditions, the centroid of the collection area has a high land value, and it would be impractical to locate a solid waste transfer station in this area. Figure 3.8 shows the effect of the distance between the potential transfer station site and the disposal site on the hauling cost.

Consider a general case in which the transfer station is located away from the hauling route between the collection area and the disposal site. Let Z be the additional distance travelled by the vehicles. The cost T, when using a transfer station, is then equal to:

$$T = B + 2AX + 2AZ + 2CY + 2CZ$$

The cost of direct hauling from the collection area to the disposal site remains the same as previously defined. The use of a transfer station is justified, if:

$$B + 2AX + 2AZ + 2CY + 2AZ < 2AX + 2AY$$

or

$$Y > (B + 2CZ + 2AZ)/(2A - 2C)$$

Again, the decision whether or not to use a transfer station is independent of the distance between the collection area and the proposed transfer station.

LEARNING ACTIVITY 3.4

Explain the role of a transfer station in solid waste management.

Note:

a) Write your answer in the space given below.

b) Check your answer with the one given at the end of this book.

3.5 WASTE COLLECTION SYSTEM DESIGN

After we identify appropriate options for collection, equipment and transfer, we must examine the various combinations of these elements to define system-wide alternatives for further analysis. Each should be evaluated for its ability to achieve the identified goals of the collection programme. Economic analysis will usually be a central focus of the system evaluation. This initial evaluation will lead to several iterations, with the differences between the alternatives under consideration becoming more narrowly focused with each round of evaluations (EPA, 1995). After comparing the alternative strategies, the various elements like crew and truck requirement, time requirement and cost involved are calculated. The various formulae used to calculate are:

(i) **Number of services/vehicle load (N):**

$$N = (C \times D)/W$$

where C = Vehicle capacity (m^3); D = Waste density (kg/m^3) and W = Waste generation/residence (kg/service).

(ii) **Time required to collect one load (E):**

$$E = N \times L$$

where L = Loading time/residence, including on-route travel.

(iii) **Number of loads/crew/day (n):**

The number of loads (n) that each crew can collect in a day can be estimated based on the workday length (t), and the time spent on administration and breaks (t_1), time for hauling and other travel (t_2) and collection route time (t_3).

- **Administrative and break time (t_1):**

  $$t_1 = A + B$$

  where A = Administrative time (i.e., for meetings, paperwork, unspecified slack time) and B = Time for breaks and lunch.

- **Hauling and other travel time (t_2):**

  $$t_2 = (n \times H) - f + G + J$$

  where n = Number of loads/crew/day; H = Time to travel to disposal site, empty truck, and return to route; f = Time to return from site to route; G = Time to travel from staging garage to route and J = Time to return from disposal site to garage.

- **Time spent on collection route (t_3):**

  $$t_3 = n \times E$$

  where variables have been previously defined.

- **Length of workday (t):**

  $$t = t_1 + t_2 + t_3$$

  where t is defined by work rules and above equations are solved to find n.

(iv) **Calculation of number of vehicles and crews (K):**

$$K = (S \times F)/(N \times n \times M)$$

where S = Total number of services in the collection area; F = Frequency of collection (numbers/week) and M = Number of workdays/week.

(v) **Calculation of annual vehicle and labour costs:**

Vehicle costs = Depreciation + Maintenance + Consumables + Overhead + License + Fees + Insurance

Labour costs = Drivers salary + Crew salaries + Fringe benefits + Indirect labour + Supplies + Overhead

✍ LEARNING ACTIVITY 3.5

Work out the time required to dispose off the waste with density 400 kg/m^3, with vehicle capacity of 12 m^3 and waste generation per residence is 2 kg/service and loading time per residence is two minutes.

Note:

a) Write your answer in the space given below.

b) Check your answer with the one given at the end of this book.

3.6 RECORD KEEPING, CONTROL, INVENTORY AND MONITORING

For effective waste collection and, indeed, SWM, we must maintain records on the quantities of wastes collected and their variation within a week, month and year, as well as on established long-term trends in solid waste generation rates and composition, sources of wastes and the personnel collecting them. Long-term trends in solid waste generation rates and composition form the basis for planning, especially in budgeting for future vehicle requirements, allocating the collection vehicles and crew, building transfer stations, acquiring strategic lands and determining disposal options. Table 3.2 contains an illustration of a checklist of factors that affect the waste collection system.

Table 3.2 Checklist of Variables Affecting Collection System

Components	*Factors to Consider*
Crew size	• labour cost • distance between containers • size and types of containers • loading accessories available in the truck • collection vehicle used
Container type	• solid wastes generation rate • density of waste generation • street width • traffic volume • collection crew configuration • standard of living
Collection accessory	• labour cost • protection of workers' health
Vehicle size/type	• street width, traffic volume • solid waste generation rates • crew size • viability of a transfer station
Collection route	• street width, traffic volume • direction of traffic flow • solid waste generation rates • spatial distribution of wastes • local topography
Transfer station	• distance between disposal site and collection area • hauling cost for small and large trucks • cost of transferring the solid wastes from small to large trucks

Source: Phelps et al., 1995

Records of personnel and quantities of wastes collected are, when maintained, useful in determining the efficiency of the personnel and in correlating waste quantities with conditions in the service area. A time keeping system at the transfer or disposal site is a key element in improving the efficiency of collection system and planning an upgraded system. The timekeeping system determines if the crew were taking long rest periods, spending time salvaging or carrying out unauthorised activities. The performance of a particular crew in terms of the quantity of solid wastes collected per day could be compared with that of another collection crew working under similar conditions.

The composition of solid wastes should be measured at least once a year for major districts and possibly once every two years in residential areas with stagnant growth rates and development. Changes in composition affect the collection equipment and configuration of the collection system is important in designing the disposal system. Changes in an energy source (such as a shift to gas or electricity from wood or charcoal for cooking and heating), reduces the ash content of wastes, making the solid waste lighter, in which case, larger containers could be used. The same line of analysis holds true in specifying the collection vehicles. Comparison of the routes taken by various crew serving a particular area helps to identify the best hauling route. Although this route may be longer, it could be more economical in terms of hauling time. However, note that the best route often changes with the season.

All these decisions should be based on reliable data, without which the waste collection system will inevitably be ineffective. Proper interpretation of monitoring data allows the authority to adapt the proposed system to actual conditions. In some instances, it also allows management to identify areas, where the design is not realistic.

✍ LEARNING ACTIVITY 3.6

Discuss the practice of record keeping and monitoring with the local civic authorities of your locality and write a short report.

Note:

a) Write your answer in the space given below.
b) Check your answer with the one given at the end of this book.

3.7 IMPLEMENTING COLLECTION AND TRANSFER SYSTEM

Implementing of collection and transfer system involves the following activities, which are important for success of the plan (EPA, 1995):

(i) **Finalising and implementing the system management plan:** For proper implementation of collection and transfer system, it is necessary to have clear organisational structures and management plans. The organisational structure should be simple, with a minimum of administrative and management layers between collection crews and top management. All workers in the department should clearly understand the department's mission and their roles. Through training, incentives and reinforcement by management, workers should be encouraged to be customer-oriented and team contributors. Feedback mechanisms must be introduced to help the crew review their performance and help managers monitoring the performance of crews, equipment, etc. It is also important to periodically review the management plans and structures, as implementation of collection services continues.

(ii) **Purchasing and managing equipment:** For purchasing equipment, most municipalities issue bid specifications. Detailed specifications include exact requirements for equipment sizes and capacities, power ratings, etc. Performance specifications often request that equipment be equivalent to certain available models and meet standards for capacity, speed, etc. Municipalities may either perform equipment maintenance themselves, contract with a local garage, or in some cases, contract with the vehicle vendor at the time of purchase. As part of the preventive maintenance programme, the collection crew should check the vehicle chassis, tyres and body daily and report any problems to maintenance managers. In addition, each vehicle should have an individual maintenance record that includes:

- preventive maintenance schedule;
- current list of specific engine;
- a description of repairs and a list containing information on the repair date, mechanic, cost, type and manufacturer of repair parts and the length of time the truck was out of service, for each maintenance event.

(iii) **Hiring and training personnel:** As in all organisations, good personnel management is essential to an efficient, high-quality waste collection system. Authorities responsible for SWM should, therefore, strive to hire and keep well-qualified personnel. The recruitment programme should assess applicants' abilities to perform the types of physical labour required for the collection, equipment and methods used. To retain employees, management should provide a safe working environment that emphasises career advancement, participatory problem solving and worker incentives. Worker incentives should be developed to recognise and reward outstanding performance by employees. Ways to accomplish motivation include merit-based compensation, awards programme and a work structure. Feedback on employee performance should be regular and frequent. Safety is especially important because waste collection employees encounter many hazards during each workday. As a result of poor safety records, insurance costs for many collection services are high. To minimise injuries, haulers should have an ongoing safety programme. This programme should outline safety procedures and ensure that all personnel are properly trained on safety issues. Haulers should develop an employee-training programme that helps employees improve and broaden the range of their job-related skills. Education should address such subjects as driving skills, first aid, safe lifting methods, identification of household hazardous wastes, avoidance of substance abuse and stress management.

(iv) **Providing public information:** Maintaining good communication with the public is important to a well-run collection system. Residents can greatly influence the performance of the collection system by co-operating in separation requirements, and by keeping undesirable materials from entering the collected waste stream. Commonly used methods of communicating information include brochures, articles in community newsletters, newspaper articles, announcements, and advertisements on radio and television, information attachments to utility bills (either printed or given separately) and school handouts. Communication materials should be used to help residents understand the community waste management challenges and the progress in meeting them. Residents should also be kept informed about issues such as the availability and costs of landfill capacity so that they develop an understanding of the issues and a desire to help meet their waste management needs.

(v) **Monitoring system cost and performance:** Collection and transfer facilities should develop and maintain an effective system for cost and performance reporting. Each collection crew should complete a daily report containing the following information:

- Total quantity hauled.
- Total distance and travel time to and from the disposal site.
- Amounts delivered to each disposal, transfer, or processing facility.
- Waiting time at sites.

- Number of loads hauled.
- Vehicle or operational problems needing attention.

Collected data should be used to forecast workloads, truck costs, identify changes in the generation of wastes and recyclables, trace the origin of problematic waste materials and evaluate crew performance. Just as the goals of a collection programme set its overall directions, a monitoring system provides the short-term feedback necessary to identify the corrections needed to achieve those goals.

✍ LEARNING ACTIVITY 3.7

Identify the activities necessary for successful implementation of collection and transfer system.

Note:

a) Write your answer in the space given below.

b) Check your answer with the one given at the end of this book.

3.8 THE CASE OF BANGALORE

In Bangalore city (India), the waste collected through street sweeping is the main system of primary collection of wastes. However, recently efforts are being made for doorstep collection of waste through NGOs (Non-Governmental Organisations) and private contractors, but only about 5% of the population is covered under this system. The waste generated by the rest is collected from either the street or the dustbins. Other details regarding the collection process in Bangalore are given below.

(i) **Waste storage:** There are about 14,000 bottomless cement bins having 0.9 metres diameter and 0.6 cubic metre storage capacity and large masonry bins for depositing wastes at a distance of about 100 to 200 metres. Besides these, there are 1500 places, where the waste is deposited but no bins are kept on these sites. Recently, metal containers have been placed and at present 55 metal containers are in the city for the storage of waste in a more hygienic manner.

(ii) **Waste collection:** The frequency of removal of wastes varies from place to place, depending on the locality. Whichever system adapted in the area needs proper planning for collection, loading, unloading and transportation from and to the point of final disposal, considering traffic constraints, peak hour traffic, etc. An optimum collection schedule requires to be worked out where the number of premises or dumps is mentioned on a daily programme sheet, to be executed by the driver or supervisor in charge of collection. At present, it is estimated that there are about 4943 hotels/ restaurants, which produce a large quantity of organic wastes in Bangalore. The silt and waste removed from drains get deposited along the roadside. The human and animal excreta also add to the mass getting deposited on streets. Mechanical sweeping or cleaning cannot work in Bangalore roads and footways because of obstructions due to the activities of hawkers, shop extensions, broken pavements, etc. Pedestrians, shops, goods vehicles carrying loose materials contribute to street littering of paper, used tickets, cigarette butts, etc., as well as vehicles dropping material during their movement (*http://stratema.sigis.net/cupum/pdf/E1.pdf*).

(iii) **Waste transportation:** Removal of garbage is a very important aspect of SWM, and the method of transportation is crucial. In essence, any breakdown in this system could create problems. Transportation implies conveyance from point of collection to the point of final disposal either directly or through a transfer system. In Bangalore, the transportation of waste is done by:

- engaging, departmentally, 82 trucks of the Corporation,
- engaging 129 vehicles, on contract, for layout and markets and 72 vehicles for transportation of waste. (In addition, the Bangalore Corporation has 13 dumper placers for transporting metallic containers of 2.5 to 3 tonnes capacity and six mini-compactors for transportation of waste.)

The clearing efficiency is 30 to 35%. The vehicles are open and the spillage of waste on the roads is a common feature. Each truck makes two trips to the dumpsite everyday. There are no transfer stations in Bangalore. The waste collected from the roads and bins is directly transported to the final disposal sites. There is no arrangement made for the primary collection of construction waste. The engineering division of the Corporation removes the unauthorised construction waste from time to time. There are 115 small and big vegetable, fruit and meat markets in the city. However, no special arrangements are made for the collection of waste from these markets.

LEARNING ACTIVITY 3.8

A collection vehicle is found to be able to service customers at a rate of two customers per minute. If actual time spent for collection is four hours per day, compute the number of customers served per day.

Calculate the number of collection vehicles a corporation ward with 10,000 houses (services) that are to be collected once per week. (The ward wants the collection on Mondays, Tuesdays, Thursdays, and Fridays, leaving Wednesdays for special projects and maintenance. A single collection vehicle can service 250 houses in a single day and still have time to take the full loads to the landfill.)

Note:

a) Write your answer in the space given below.

b) Check your answer with the one given at the end of this book.

SUMMARY

In this Unit, our focus was on waste collection systems. We discussed in detail the various components of waste collection system including collection frequency, storage containers, collection route and transfer station. We also discussed the viability of establishing and maintaining transfer stations. During the course of our discussion on the movement of collection vehicles and crew, we brought out the usefulness of the MTM technique to achieve efficiency in vehicle and crew movement. We also discussed the design, operation and implementation aspects of waste collection systems. We closed the Unit with the case study of Bangalore.

SUGGESTED READINGS

Tchobanoglous, G., Theisen, H. and Eliassan, R. 1977. Solid Wastes—Engineering Principles and Management Issues. McGraw-Hill Series, Kogakusha.

UNCHS. 1994. Promotion of Solid Waste Recycling and Reuse in the Developing Countries of Asia—A Reference Handbook for Trainers, Kenya.

Arne Vesilind, William Worrel and Reinhart Debra. 2002. Solid Waste Engineering. Thomson Brooks/Cole, Singapore.

Tchobanoglous, G., Theisen, H. and Samuel A. Vigil, 1993. Integrated Solid Waste Management, McGraw-Hill, Inc., New Delhi.

Flinotoff, R.1984. Management of Solid Wastes in Developing Countries, WHO Regional Publications, South East Asia Series No 1.

http://ces.iisc.ernet.in/energy/SWMTR/TR85.html

REFERENCES

Ali, S.M., Cotton, A.P. and Westlake, K. 1999. Down to Earth: Solid Waste Disposal for Low-Income Countries, WEDC, Loughborough University, UK.

Phelps, H.O., Heinke, G.W., Jonker, J.F., Ouano, E.A.R. and Vandecasteele, C. 1995. Management of Solid Wastes, UNESCO, Paris.

UNEP.1996. International Source Book on Environmentally Sound Technologies for Municipal Solid Waste Management (6), IETC, Osaka/Shiga.

US Environmental Protection Agency. 1989. *Decision-Maker's Guide to Solid Waste Management, Vol I*, Washington.

US Environmental Protection Agency. 1995. *Decision-Maker's Guide to Solid Waste Management, Vol II*, Washington.

http://stratema.sigis.net/cupum/pdf/E1.pdf

UNIT 4

Waste Disposal

STRUCTURE

OVERVIEW

In Unit 3, we examined waste collection system that subsumes activities such as waste collection, storage and transportation. As a logical sequence, the wastes collected should properly be disposed off to avoid health and environmental hazards, and this is the focus of Unit 4. We will begin this Unit by explaining the key issues in waste disposal followed by disposal options and selection criteria. Then, we will discuss the engineered waste disposal option in detail and explain the principles, processes, environmental concerns and issues related to its design, construction and monitoring. Finally, we will present a case study of disposal option.

LEARNING OBJECTIVES

After completing this Unit, you should be able to:

- assess key issues associated with waste disposal;
- evaluate the various options for disposal of wastes and their selection criteria;
- explain the design, operation and maintenance of sanitary landfill;
- determine the most viable disposal option for your locality.

4.1 KEY ISSUES IN WASTE DISPOSAL

Let us first get one thing very clear: there is no option but to dispose off wastes. Disposal is the final element in the SWM system. It is the ultimate fate of all solid wastes, be they residential wastes collected and transported directly to a landfill site, semisolid waste (sludge) from municipal and industrial treatment plants, incinerator residue, compost or other substances from various solid waste processing plants that are of no further use to society. It is, therefore, imperative to have a proper plan in place for safe disposal of solid wastes, which involves appropriate handling of residual matter after solid wastes have been processed and the recovery of conversion products/energy has been achieved. It follows that an efficient SWM system must provide an environmentally sound disposal option for waste that cannot be reduced, recycled, composted, combusted, or processed further (Ali et al., 1999).

However, these days indiscriminate disposal of wastes in many regions is very common, giving rise to such problems as:

- health hazards (e.g., residents in the vicinity of wastes inhale dust and smoke when the wastes are burnt; workers and rag pickers come into direct contact with wastes, etc.);
- pollution due to smoke;
- pollution from waste leachate and gas;
- blockage of open drains and sewers.

Clearly, safe disposal of solid wastes is important for safeguarding both public health and the environment.

Issues to be Overcome

To achieve effective waste disposal, we must overcome the following constraints:

(i) **Municipal capacities:** With the increasing volume of waste generation, collection of wastes gets more attention than disposal. Furthermore, in India, only a few municipalities seem to have the required experience or capacity for controlled disposal. Some municipalities may have identified disposal sites but still only few may actively manage them. In some places, contracting out waste disposal is seen as a solution. But, municipalities are not equipped to deal with the problems associated with it, such as issues of privatisation and monitoring of the contract.

(ii) **Political commitment:** SWM is more than a technical issue, as any successful programme needs effective political and governmental support. This is rarely a priority of government authorities, unless there is a strong and active public interest as well as international interventions.

(iii) **Finance and cost recovery:** Development of a sanitary landfill site represents a major investment and it generally receives less priority over other resource demands. And, even when establishment costs are secured for a disposal site, recurrent costs to maintain it always pose problems.

(iv) **Technical guidelines:** Standards established for waste disposal in one country need not necessarily be appropriate for another, due to reasons such as climatic conditions, resources availability, institutional infrastructure, socio-cultural values, etc. In the absence of adequate data and/or the means of collecting/acquiring it, officials often struggle to plan a safe and economically viable disposal option.

(v) **Institutional role and responsibility:** A disposal site may be located outside the boundary of a town and may serve more than one town. This necessitates the co-ordination of all authorities concerned, and the roles and responsibilities of different departments need to be clearly defined and accepted by all concerned.

(vi) **Location:** The accessibility of a disposal site, especially its distance from town, is an important factor in site selection, especially when staff and public do not have a strong incentive to use it, when compared to indiscriminate dumping. Site selection is perhaps the most difficult stage in the development of suitable disposal option.

✍ **LEARNING ACTIVITY 4.1**

Identify waste disposal constraints faced by municipal authorities of your locality.

Note:

a) Write your answer in the space given below.

b) Check your answer with the one given at the end of this book.

4.2 DISPOSAL OPTIONS AND SELECTION CRITERIA

The most common disposal option practised currently in many countries is either uncontrolled dumping or dumping with moderate control. The environmental costs of uncontrolled dumping include breeding of disease causing vectors (e.g., flies, mosquitoes and rodents), pollution, odour and smoke. In Subsection 4.2.1, we will discuss some of the options available for waste disposal. However, it should be noted that the option selected for waste disposal must mesh with the existing socio-cultural milieu, infrastructure, etc., and this we will discuss in Subsection 4.2.2.

4.2.1 Disposal Options

In this subsection, we will touch upon some of the options available for waste disposal, and in that respect, we will consider the following:

(i) **Uncontrolled dumping or non-engineered disposal:** As mentioned, this is the most common method being practised in many parts of the world, and India is no exception. In this method, wastes are dumped at a designated site without any environmental control. They tend to remain there for a long period of time, pose health risks and cause environmental degradation. Due to the adverse health and environmental impact associated with it, the non-engineered disposal is not considered a viable and safe option.

(ii) **Sanitary landfill:** Unlike the non-engineered disposal, sanitary landfill is a fully engineered disposal option in that the selected location or wasteland is carefully engineered in advance before it is pressed into service. Operators of sanitary landfills can minimise the effects of leachate (i.e., polluted water which flows from a landfill) and gas production through proper site selection, preparation and management. This particular option of waste disposal is suitable when the land is available at an affordable price, and adequate workforce and technical resources are available to operate and manage the site. We will discuss this option in detail in Section 4.3.

(iii) **Composting:** This is a biological process of decomposition in which organisms, under controlled conditions of ventilation, temperature and moisture, convert the organic portion of solid waste into humus-like material. If this process is carried out effectively, what we get as the final product is a stable, odour-free soil conditioner. Generally, the option of composting is considered, when a considerable amount of biodegradable waste is available in the waste stream and there is use or market for composts. Composting can be either centralised or small-scale. Centralised composting plants are possible, if adequate skilled workforce and equipments are available. And, small-scale composting practices can be effective at household level, but this needs public awareness. We will discuss composting processes, methods, technologies and environmental consequences in detail in Unit 7.

(iv) **Incineration:** This refers to the controlled burning of wastes, at a high temperature (roughly 1200–1500 °C), which sterilises and stabilises the waste in addition to reducing its volume. In the process, most of the combustible materials (i.e., self-sustaining combustible matter, which saves the energy needed to maintain the combustion) such as paper or plastics get converted into carbon dioxide and ash. Incineration may be used as a disposal option, when land filling is not possible and the waste composition is highly combustible. An appropriate technology, infrastructure and skilled workforce are required to operate and maintain the plant. We will discuss in detail the process, technology and environmental concerns of incineration, which is generally limited to hospital and other biological wastes, in Unit 8.

(v) **Gasification:** This is the partial combustion of carbonaceous material (through combustion) at high temperature (roughly 1000 °C) forming a gas, comprising mainly carbon dioxide, carbon monoxide, nitrogen, hydrogen, water vapour and methane, which can be used as fuel. We will discuss the aspects of energy recovery, including gasification and refuse-derived fuel (RDF), described below, in Unit 8.

(vi) **Refuse-derived fuel (RDF):** This is the combustible part of raw waste, separated for burning as fuel. Various physical processes such as screening, size reduction, magnetic separation, etc., are used to separate the combustibles (see Unit 8 for details).

(vii) **Pyrolysis:** This is the thermal degradation of carbonaceous material to gaseous, liquid and solid fraction in the absence of oxygen. This occurs at a temperature between 200 and 900 °C The product of pyrolysis is a gas of relatively high calorific value of 20,000 joules per gram with oils, tars and solid burned residue (Ali et al., 1999).

Relative Merits of Some Options

Having touched upon several disposal options, let us now present the merits and demerits of some of them in Table 4.1.

Table 4.1 Relative Merits of Disposal Options

Disposal Option → / *↓ Sustainability Indicator*	*Non-engineered Disposal*	*Sanitary Landfill*	*Composting*	*Incineration*
Volume reduction	×	×	×	√
Expensive	×	√	√	√
Long-term maintenance	√	√	×	×
By-product recovery	×	√	√	√
Adaptability to all wastes	√	√	×	×
Adverse environmental effect	√	√	×	√

Source: http//ces.iisc.ernet.in/energy/SWMTR/TR85.html

4.2.2 Selection Criteria

With the help of proper frameworks and sub-frameworks, we can assess the effectiveness of each of the waste disposal options (see Subsection 4.2.1). While a framework represents an aid to decision-making and helps to ensure that the key issues are considered, a sub-framework explains how and why the necessary information should be obtained (Ali et al., 1999). A framework contains a list of issues and questions pertaining to the technical, institutional, financial, social and environmental features of a waste disposal system to assess the capacity of a disposal option to meet the requirements. For example, an appraisal of waste disposal option must include the following:

(i) **Technical:** This feature, involving efficient and effective operation of the technology being used, evaluates the following components of a SWM system:

- composition of wastes, e.g., type, characteristics and quantity.
- existing practices, e.g., collection, transport, and recycling process.
- siting, e.g., location of disposal site, engineering material, etc.
- technology, e.g., operation, maintenance, technical support, etc.
- impact, e.g., anticipated by-product, requirement for their treatment and disposal, etc.

(ii) **Institutional:** This involves the ability and willingness of responsible agencies to operate and manage the system by evaluating the following:

- structures, roles and responsibilities, e.g., current institutional frameworks.
- operational capacity, e.g., municipal capacities, local experience and staff training.
- incentives, e.g., management improvement and waste disposal practices.
- innovation and partnership.

(iii) **Financial:** This assesses the ability to finance the implementation, operation and maintenance of the system by evaluating the following:

- financing and cost recovery, e.g., willingness to raise finance for waste management.
- current revenue and expenditure on waste management.
- potential need for external finance for capital cost.

(iv) **Social:** This helps in avoiding adverse social impact by evaluating the following:

- waste picking, which has an impact on livelihood and access to waste pickers.
- health and income implication.
- public opinions on the existing and proposed system.

(v) **Environmental:** This means setting up an environment friendly disposal system by evaluating the following:

- initial environmental risks, i.e., impact of existing and proposed disposal option.
- long-term environmental risks, i.e., long-term implication (future impacts).

LEARNING ACTIVITY 4.2

Considering the selection criteria, suggest the best disposal option for your locality.

Note:

a) Write your answer in the space given below.

b) Check your answer with the one given at the end of this book.

We touched upon the various disposal options alongside the selection criteria for disposal options. One such option we mentioned is engineered disposal, often referred to as sanitary landfill. Although in several countries, uncontrolled dumping is still the most prevalent disposal option, sanitary landfill is gradually taking precedence as the ill effects of uncontrolled dumping are increasing. This being so, we will discuss the principle, processes, design, construction and monitoring aspects of sanitary landfill next.

4.3 SANITARY LANDFILL

The term landfill generally refers to an engineered deposit of wastes either in pits/trenches or on the surface. And, a sanitary landfill is essentially a landfill, where proper mechanisms are available to control the environmental risks associated with the disposal of wastes and to make available the land, subsequent to disposal, for other purposes. However, you must note that a landfill need not necessarily be an engineered site, when the waste is largely inert at final disposal, as in rural areas, where wastes contain a large proportion of soil and dirt. This practice is generally designated as non-engineered disposal method. When compared to uncontrolled dumping, engineered landfills are more likely to have pre-planned installations, environmental monitoring, and organised and trained workforce. Sanitary landfill implementation, therefore, requires careful site selection, preparation and management.

The four minimum requirements you need to consider for a sanitary landfill are:

(i) full or partial hydrological isolation;
(ii) formal engineering preparation;
(iii) permanent control; and
(iv) planned waste emplacement and covering.

Against this background, let us now discuss the principles, processes and operation of sanitary landfills.

4.3.1 Principle

The purpose of land filling is to bury or alter the chemical composition of the wastes so that they do not pose any threat to the environment or public health. Landfills are not homogeneous and are usually made up of cells in which a discrete volume of waste is kept isolated from adjacent waste cells by a suitable barrier. The barriers between cells generally consist of a layer of natural soil (i.e., clay), which restricts downward or lateral escape of the waste constituents or leachate.

Land filling relies on containment rather than treatment (for control) of wastes. If properly executed, it is a safer and cheaper method than incineration (see Unit 8). An environmentally sound sanitary landfill comprises appropriate liners for protection of the groundwater (from contaminated leachate), run-off controls, leachate collection and treatment, monitoring wells and appropriate final cover design (Phelps, 1995). Figure 4.1 gives a schematic layout of sanitary landfill along with its various components:

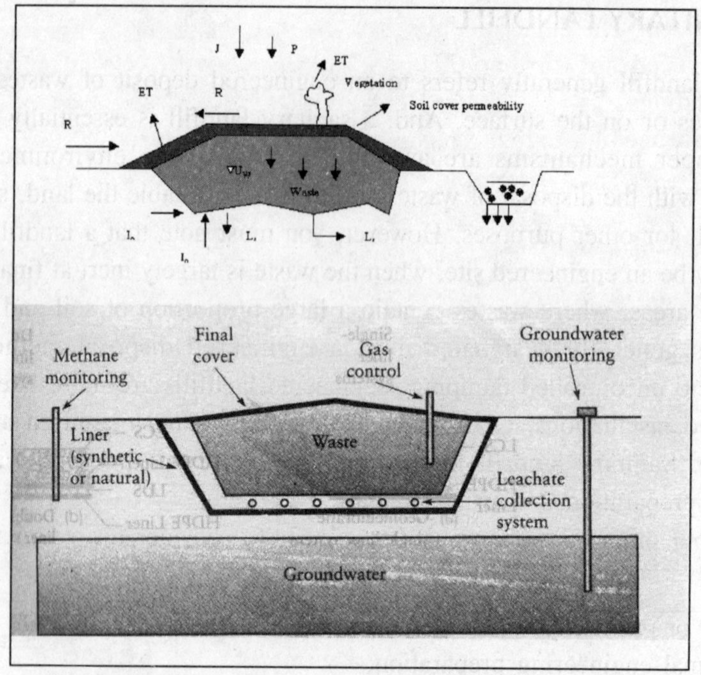

Figure 4.1 Schematic Layout of Sanitary Landfill

P: precipitation; J: irrigation or leachate recirculation; R: surface runoff; R: runoff from external areas; ET: actual evapotranspiration;* $P_i = P+J+R^* - R - ET \pm \Delta Us$*; Us: water content in soil; Uw: water content in waste; S: water added in sludge disposal; b: water production (if* >0*) or consumption (if* <0*) caused by biological degradation of organic matter; Is/Ig: water from natural aquifers;* $L = Pi + S + Ig + b \pm \Delta Uw$*; L: total leachate production; Li: infiltration into aquifers; Lr: leachate collected by drains.*

Before we take up landfill processes for discussion in Subsection 4.3.2, let us touch upon the phases in the life cycle of a landfill, and these are:

- **Planning phase**: This typically involves preliminary hydro-geological and geo-technical site investigations as a basis for actual design.
- **Construction phase**: This involves earthworks, road and facility construction and preparation (liners and drains) of the fill area.
- **Operation phase (5–20 years):** This phase has a high intensity of traffic, work at the front of the fill, operation of environmental installations and completion of finished sections.
- **Completed phase (20–100 years):** This phase involves the termination of the actual filling to the time when the environmental installations need no longer

be operated. The emissions may have by then decreased to a level where they do not need any further treatment and can be discharged freely into the surroundings.

- **Final storage phase:** In this phase, the landfill is integrated into the surroundings for other purposes, and no longer needs special attention.

4.3.2 Landfill Processes

The feasibility of land disposal of solid wastes depends on factors such as the type, quantity and characteristics of wastes, the prevailing laws and regulations, and soil and site characteristics. Let us now explain some of these processes.

(i) **Site selection process:** This requires the development of a working plan – a plan, or a series of plans, outlining the development and descriptions of site location, operation, engineering and site restoration.

(ii) **Settling process:** The waste body of a landfill undergoes different stages of settling or deformation. Figure 4.2 below illustrates these stages.

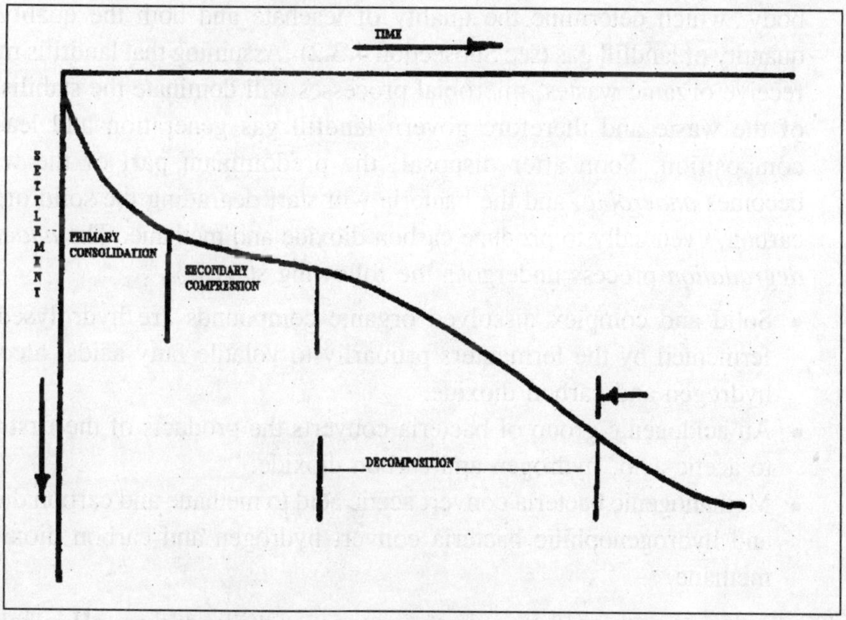

Figure 4.2 Settling Processes in Landfill

The three stages shown in the figure above are described below:

- **Primary consolidation:** During this stage, a substantial amount of settling occurs. This settlement is caused by the weight of the waste layers. The movement of trucks, bulldozers or mechanical compactors will also enhance this process. After this primary consolidation, or short-term deformation, stage, *aerobic degradation* processes occur.

- **Secondary compression:** During this stage, the rate of settling is much lower than that in the primary consolidation stage, as the settling occurs through compression, which cannot be enhanced.
- **Decomposition:** During the degradation processes, organic material is converted into gas and leachate. The settling rate during this stage increases compared to the secondary compression stage, and continues until all decomposable organic matter is degraded. The settling rate, however, gradually decreases with the passage of time.

To appropriately design protective liners, and gas and leachate collection systems, it is, therefore, necessary to have a proper knowledge of the settling process of wastes.

(iii) **Microbial degradation process:** The microbial degradation process is the most important biological process occurring in a landfill. These processes induce changes in the chemical and physical environment within the waste body, which determine the quality of leachate and both the quality and quantity of landfill gas (see Subsection 4.3.2). Assuming that landfills mostly receive organic wastes, microbial processes will dominate the stabilisation of the waste and therefore govern landfill gas generation and leachate composition. Soon after disposal, the predominant part of the wastes becomes *anaerobic*, and the bacteria will start degrading the solid organic carbon, eventually to produce carbon dioxide and methane. The *anaerobic degradation* process undergoes the following stages:

- Solid and complex dissolved organic compounds are hydrolysed and fermented by the fermenters primarily to volatile fatty acids, alcohols, hydrogen and carbon dioxide.
- An acidogenic group of bacteria converts the products of the first stage to acetic acid, hydrogen and carbon dioxide.
- Methanogenic bacteria convert acetic acid to methane and carbon dioxide and hydrogenophilic bacteria convert hydrogen and carbon dioxide to methane.

The biotic factors that affect methane formation in the landfill are pH, alkalinity, nutrients, temperature, oxygen and moisture content.

Enhancement of Degradation

Enhancement of the degradation processes in landfills will result in a faster stabilisation of the waste in the landfill, which enhances gas production, and we can achieve this by:

- **Adding partly composted waste:** As the readily degradable organic matter has already been decomposed aerobically, the rapid acid production phase is overcome, and the balance of acid and methane production bacteria can develop earlier and the consequent dilution effect lowers the organic acid concentration.
- **Recirculating leachate:** This may have positive effects since a slow increase in moisture will cause a long period of gas production. During warmer periods, recirculated leachate will evaporate, resulting in lower amounts of excess leachate.

✍ LEARNING ACTIVITY 4.3

Do you think a sanitary landfill is possible to manage wastes in your locality? List at least three reasons to support your answer.

Note:

a) Write your answer in the space given below.
b) Check your answer with the one given at the end of this book.

We mentioned earlier that microbial degradation of waste under anaerobic conditions induces gas emission and leachate formation. We will explain this further, next.

Landfill Gas and Leachate

Leachate and landfill gas comprise the major hazards associated with a landfill. While leachate may contaminate the surrounding land and water, landfill gas can be toxic and lead to global warming and explosion leading to human catastrophe (Phelps, 1995). (Note that global warming, also known as greenhouse effect, refers to the warming of the earth's atmosphere by the accumulation of gases (e.g., methane, carbon dioxide and chlorofluorocarbons) that absorbs reflected solar radiation.) The factors, which affect the production of leachate and landfill gas, are the following:

- **Nature of waste:** The deposition of waste containing biodegradable matter invariably leads to the production of gas and leachate, and the amount depends on the content of biodegradable material in the waste.
- **Moisture content:** Most micro-organisms require a minimum of approximately 12% (by weight) moisture for growth, and thus the moisture content of landfill waste is an important factor in determining the amount and extent of leachate and gas production.
- **pH:** The methanogenic bacteria within a landfill produce methane gas, which will grow only at low pH range around neutrality.
- **Particle size and density:** The size of waste particle affects the density that can be achieved upon compaction and affects the surface area and hence volume. Both affect moisture absorption and therefore are potential for biological degradation.
- **Temperature:** An increase in temperature tends to increase gas production. The temperature affects the microbial activity to the extent that it is possible to segregate bacteria, according to their optimum temperature operating conditions.

Note that the composition of waste, which varies with region and climate (season), determines the variation in pollution potential. Carbohydrates comprise a large percentage of biodegradable matter within municipal waste, the overall breakdown of which can be represented by the following equation:

$$\underset{\text{(Carbohydrate)}}{C_6H_{12}O_6} \xrightarrow{\text{(Bacteria)}} \underset{\text{(Methane)}}{CH_4} + \underset{\text{(Carbon dioxide)}}{CO_2} + \text{Biomass} + \text{Heat}$$

Let us now discuss landfill leachate and gas emission in detail along with their composition and adverse effects.

4.4 LANDFILL GAS EMISSION

Landfill gas contains a high percentage of methane due to the anaerobic decomposition of organic matter, which can be utilised as a source of energy. In Subsections 4.4.1 to 4.4.4, we will explain the composition and properties, risks, migration and control of landfill gas.

4.4.1 Composition and Properties

We can predict the amount and composition of the gas generated for different substrates, depending on the general anaerobic decomposition of wastes added. Climatic and environmental conditions also influence gas composition. Due to the heterogeneous nature of the landfill, some acid-phase anaerobic decomposition occurs along with the methanogenic decomposition. Since aerobic and acid-phase degradation give rise to carbon dioxide and not methane, there may be a higher carbon dioxide content in the gas generated than what would otherwise be expected. Furthermore, depending on the moisture distribution, some carbon dioxide goes into solution. This may appear to increase (artificially) the methane content of the gas measured in the landfill. A typical landfill gas contains a number of components such as the following, which tend to occur within a characteristic range:

- **Methane:** This is a colourless, odourless and flammable gas with a density lighter than air, typically making up 50–60% of the landfill gas.
- **Carbon dioxide:** This is a colourless, odourless and non-inflammable gas that is denser than air, typically accounting for 30–40%.
- **Oxygen:** The flammability of methane depends on the percentage of oxygen. It is, therefore, important to control oxygen levels, where gas abstraction is undertaken.
- **Nitrogen:** This is essentially inert and will have little effect, except to modify the explosive range of methane.

It is difficult to convert the amount of gas measured to the maximum landfill gas production value because gas is withdrawn from a small part of the landfill only, referred to as *zone of influence* during measurement. In other words, it is very difficult to determine this zone and relate it to the whole landfill area.

4.4.2 Hazards

Landfill gas consists of a mixture of flammable, asphyxiating and noxious gases and may be hazardous to health and safety, and hence the need for precautions. Some of the major hazards are listed below:

- **Explosion and fire:** Methane is flammable in air within the range of 5–15% by volume, while hydrogen is flammable within the range of 4.1–7.5% (in the presence of oxygen) and potentially explosive. Fire, occurring within the waste, can be difficult to extinguish and can lead to unpredictable and uncontrolled subsidence as well as production of smoke and toxic fumes.
- **Trace components:** These comprise mostly alkanes and alkenes, and their oxidation products such as aldehydes, alcohols and esters. Many of them are

recognised as toxicants, when present in air at concentrations above occupational exposure standards.

- **Global warming:** Known also as greenhouse effect, it is the warming of the earth's atmosphere by the accumulation of gases (methane, carbon dioxide and chlorofluorocarbons) that absorbs reflected solar radiation.

4.4.3 Migration

During landfill development, most of the gas produced is vented to the atmosphere, provided the permeable intermediate cover has been used. While biological and chemical processes affect gas composition through methane oxidation, which converts methane to carbon dioxide, physical factors affect gas migration. The physical factors that affect gas migration include:

- **Environmental conditions:** These affect the rate of degradation and gas pressure build up.
- **Geophysical conditions:** These affect migration pathways. In the presence of fractured geological strata or a mineshaft, the gas may travel large distances, unless restricted by the water table.
- **Climatic conditions:** Falling atmospheric pressure, rainfall and water infiltration rate affect landfill gas migration.

The proportion of void space in the ground, rather than permeability, determines the variability of gas emission. If the escape of landfill gas is controlled and proper extraction system is designed, this gas can be utilised as a source of energy. If landfill gas is not utilised, it should be burnt by means of flaring. However, landfill gas utilisation can save on the use of fossil fuels since its heating value is approximately 6 kWh/m^3 and can be utilised in internal combustion engines for production of electricity and heat.

It is important that landfill gas is extracted during the operation phase. It is extracted out of the landfill by means of gas wells, which are normally drilled by auger and are driven into the landfill at a spacing of 40–70 m. In addition, horizontal systems can be installed during operation of the landfill. The gas wells consist mainly of perforated plastic pipes surrounded by coarse gravel and are connected with the gas transportation pipe with flexible tubing. The vacuum necessary for gas extraction and transportation is created by means of a blower. The most important factors influencing planning and construction of landfill gas extraction systems are settling of waste, water tables in landfills and gas quality.

4.4.4 Control

To control gas emission, it is necessary to control the following:

- waste inputs (i.e., restrict the amount of organic waste).

- processes within the waste (i.e., minimise moisture content to limit gas production).
- migration process (i.e., provide physical barriers or vents to remove the gas from the site and reduce gas pressure). Note that since gas migration cannot be easily prevented, removal is often the preferred option. This is done by using vents (extraction wells) within the waste or stone filled vents, which are often placed around the periphery of the landfill site.

LEARNING ACTIVITY 4.4

Identify adverse effects of landfill gas and list appropriate control measures.

Note:

a) Write your answer in the space given below.

b) Check your answer with the one given at the end of this book.

4.5 LEACHATE FORMATION

Leachate can pollute both groundwater and surface water supplies. The degree of pollution will depend on local geology and hydrogeology, nature of waste and the proximity of susceptible receptors. Once groundwater is contaminated, it is very costly to clean it up. Landfills, therefore, undergo siting, design and construction procedures that control leachate migration.

4.5.1 Composition and Properties

Leachate comprises soluble components of waste and its degradation products enter water, as it percolates through the landfill. The amount of leachate generated depends on:

- water availability;
- landfill surface condition;
- refuse state;
- condition of surrounding strata.

The major factor, i.e., water availability, is affected by precipitation, surface runoff, waste decomposition and liquid waste disposal. The water balance equation for landfill requires negative or zero ("*Lo*") so that no excess leachate is produced. This is calculated using the following formula:

$$Lo = I - E - aW$$

i.e.

$$I - E < aW$$

where Lo = free leachate retained at site (equivalent to leachate production minus leachate leaving the site); I = total liquid input; E = evapotranspiration losses; a = absorption capacity of waste; and W = weight of waste disposed.

Common toxic components in leachate are ammonia and heavy metals, which can be hazardous even at low levels, if they accumulate in the food chain. The presence of ammoniacal nitrogen means that leachate often has to be treated off-site before being discharged to a sewer, since there is no natural bio-chemical path for its removal (Ali et al., 1995). Leachate composition varies with time and location. Table 4.2 shows a typical leachate properties and composition at various stages of waste decomposition:

Table 4.2 Properties and Composition of Leachate at Various Stages of Decomposition (mg/l except pH)

Components	*Fresh wastes*	*Aged wastes*	*Wastes with high moisture*
pH	6.2	7.5	8.0
COD	23800	1160	1500
BOD	11900	260	500
TOC	8000	465	450
Volatile acid (as C)	5688	5	12
NH_3-N	790	370	1000
NO_3-N	3	1	1.0

(Contd.)

Table 4.2 (*Contd.*)

Ortho-P	0.73	1.4	1.0
Cl	1315	2080	1390
Na	9601	300	1900
Mg	252	185	186
K	780	590	570
Ca	1820	250	158
Mn	27	2.1	0.05
Fe	540	23	2.0
Cu	0.12	0.03	–
Zn	21.5	0.4	0.5
Pb	0.40	0.14	–

Source: Ali et al., 1995

4.5.2 Leachate Migration

It is generally difficult to predict the movement of escaped leachate accurately. The main controlling factors are the surrounding geology and hydrogeology. Escape to surface water may be relatively easy to control, but if it escapes to groundwater sources, it can be very difficult both to control and clean up. The degree of groundwater contamination is affected by physical, chemical and biological actions. The relative importance of each process may change, however, if the leachate moves from the landfill to the sub-surface region.

4.5.3 Control

The best way to control leachate is through prevention, which should be integral to the site design. In most cases, it is necessary to control liquid access, collection and treatment, all of which can be done using the following landfill liners:

- **Natural liners:** These refer to compacted clay or shale, bitumen or soil sealants, etc., and are generally less permeable, resistant to chemical attack and have good sorption properties. They generally do not act as true containment barriers, because sometimes leachate migrates through them.
- **Synthetic (geo-membrane) liners:** These are typically made up of high or medium density polyethylene and are generally less permeable, easy to install, relatively strong and have good deformation characteristics. They sometimes expand or shrink according to temperature and age.

Note that natural and geo-membrane liners are often combined to enhance the overall efficiency of the containment system.

4.5.4 Treatment

Concentrations of various substances occurring in leachate are too high to be discharged to surface water or into a sewer system. These concentrations, therefore, have to be reduced by removal, treatment or both. The various treatments of leachate include:

- **Leachate recirculation**: It is one of the simplest forms of treatment. Recirculation of leachate reduces the hazardous nature of leachate and helps wet the waste, increasing its potential for biological degradation.
- **Biological treatment:** This removes BOD, ammonia and suspended solids. Leachate from land filled waste can be readily degraded by biological means, due to high content of volatile fatty acids (VFAs). The common methods are aerated lagoons (i.e., special devices which enhance the aerobic processes of degradation of organic substances over the entire depth of the tank) and activated sludge process, which differs from aerated lagoons in that discharged sludge is recirculated and is often used for BOD and ammonia removal. While under conditions of low COD, rotating biological contactors (i.e., biomass is brought into contact with circular blades fixed to a common axle which is rotated) are very effective in removing ammonia. In an anaerobic treatment system, complex organic molecules are fermented in filter. The common types are anaerobic filters, anaerobic lagoon and digesters.

> ✍ **LEARNING ACTIVITY 4.5**
>
> State how you would control and treat leachate.
>
> **Note**:
>
> a) Write your answer in the space given below.
>
> b) Check your answer with the one given at the end of this book.
>
> ____
>
> ____
>
> ____
>
> ____
>
> ____
>
> ____
>
> ____
>
> ____
>
> ____

- **Physicochemical treatment:** After biological degradation, effluents still contain significant concentrations of different substances. Physicochemical treatment processes could be installed to improve the leachate effluent quality. Some of these processes are flocculation-precipitation. (Note that addition of chemicals to the water attracts the metal by floc formation.) Separation of the floc from water takes place by sedimentation, adsorption and reverse osmosis.

4.6 ENVIRONMENTAL EFFECTS OF LANDFILL

The environmental effects of a landfill include wind-blown litter and dust, noise, obnoxious odour, vermin and insects attracted by the waste, surface runoff and inaesthetic conditions. Gas and leachate problems also arise during the operation phase and require significant environmental controls. In what follows, we will describe some of the major environmental effects below:

(i) Wind-blown litter and dust are continuous problems of the ongoing landfill operation and a nuisance to the neighbourhood. Covering the waste cells with soil and spraying water on dirt roads and waste in dry periods, in combination with fencing and movable screens, may minimise the problem of wind-blown litter and dust. However, note that the problem will remain at the tipping front of the landfill.

(ii) Movement of waste collection vehicles, emptying of wastes from them, compactors, earthmoving equipment, etc., produce noise. Improving the technical capability of the equipment, surrounding the fill area with soil embankments and plantations, limiting the working hours and appropriately training the workforce will help minimise noise pollution.

(iii) Birds (e.g., scavengers), vermin, insects and animals are attracted to the landfill for feeding and breeding. Since many of these may act as disease vectors, their presence is a potential health problem.

(iv) Surface run-off, which has been in contact with the land filled waste, may be a problem in areas of intense rainfall. If not controlled, heavily polluted run-off may enter directly into creeks and streams. Careful design and maintenance of surface drains and ditches, together with a final soil cover on completed landfill sections, can help eliminate this problem.

(v) An operating landfill, where equipment and waste are exposed, appears inaesthetic. This problem may be reduced by careful design of screening soil embankments, plantings, rapid covering and re-vegetation of filled sections.

(vi) Gas released, as a result of degradation or volatilisation of waste components, causes odour, flammability, health problems and damage of the vegetation (due to oxygen depletion in the root zone). The measures to control this include liners, soil covers, passive venting or active extraction of gas for treatment before discharge into the atmosphere.

(vii) Polluted leachate appears shortly after disposal of the waste. This may cause groundwater pollution and pollution of streams through sub-surface migration. Liners, drainage collection, treatment of leachate, and groundwater and downstream water quality monitoring are necessary to control this problem.

Figure 4.3 gives a summary of the environmental emissions from a sanitary landfill.

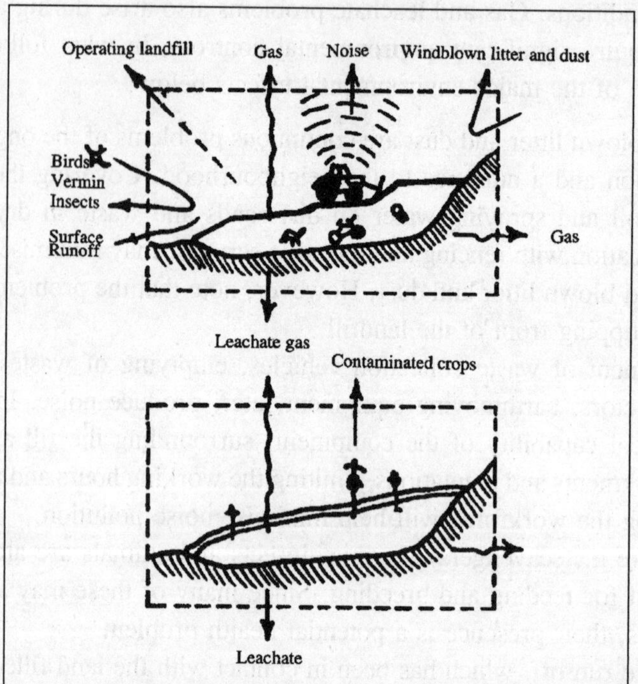

Figure 4.3 Environmental Emissions from a Sanitary Landfill

Besides the emissions shown in Fig. 4.3, incidental events such as flooding, fires, landslides and earthquakes result in severe environmental impacts, and may require preventive measures with respect to landfill site selection, design and operation. In short, to minimise adverse environmental impacts due to sanitary landfill, proper attention must be paid to the environmental aspects at all stages and phases of landfill management, viz., site selection, design, construction, operation and maintenance (Ali et al., 1995).

LEARNING ACTIVITY 4.6

Compare sanitary landfill and uncontrolled dumping from the point of view of public health and the environment.

Note:

a) Write your answer in the space given below.

b) Check your answer with the one given at the end of this book.

4.7 LANDFILL OPERATION ISSUES

Once a potential site has been identified/selected, an assessment of design aspects, including costs for civil works, begins. Important issues to be looked into in this regard are land requirements, types of wastes that are handled, evaluation of seepage potential, design of drainage and seepage control facilities, development of a general operation plan, design of solid waste filling plan and determination of equipment requirements.

With this in view, we will discuss some important factors required for successful implementation and operation of a sanitary landfill in Subsections 4.7.1 to 4.7.3.

4.7.1 Design and Construction

The design and construction process involves site infrastructure, i.e., the position of the buildings, roads and facilities that are necessary to the efficient running of the site and site engineering, i.e., the basic engineering works needed to shape the site for the reception of wastes and to meet the technical requirements of the working plan (Phelps et al., 1995). At the outset, however, the potential operator and the licensing authority should agree upon a working plan for the landfill. The disposal license includes the design, earthworks and procedures in the working plan.

What are the processes involved in design and construction? We will study these below:

(i) **Site infrastructure:** The size, type and number of buildings required at a landfill depend on factors such as the level of waste input, the expected life of the site and environmental factors. Depending on the size and complexity of the landfill, buildings range from single portable cabins to big complexes. However, certain aspects such as the following are common:

- need to comply with planning, building, fire, health and safety regulations and controls;
- security and resistance to vandalism;
- durability of service and the possible need to relocate accommodation during the lifetime of the site operations;
- ease of cleaning and maintenance;
- availability of services such as electricity, water, drainage and telecommunication.

Paying some attention to the appearance of the site entrance is necessary, as it influences the perception of the public about the landfill site. All landfill sites need to control and keep records of vehicles entering and leaving the site, and have a weighbridge to record waste input data, which can be analysed by a site control office. Note that at small sites, the site control office can be accommodated at the site itself.

(ii) **Earthworks:** Various features of landfill operations may require substantial earthworks, and therefore, the working plan must include earthworks to be carried out before wastes can be deposited. Details about earthworks gain significance, if artificial liners are to be installed, which involves grading the base and sides of the site (including construction of 25 slopes to drain leachate to the collection areas) and the formation of embankments. Material may also have to be placed in stockpiles for later use at the site. The cell method of operation requires the construction of cell walls. At some sites, it may be necessary to construct earth banks around the site perimeter to screen the landfill operations from the public. Trees or shrubs may then be planted on the banks to enhance the screening effect. The construction of roads leading to disposal sites also involves earthworks.

(iii) **Lining landfill sites:** Where the use of a liner is envisaged, the suitability of a site for lining should be evaluated at the site investigation stage. However, they should not be installed, until the site has been properly prepared. The area to be lined should be free of objects likely to cause physical damage to the liner, such as vegetation and hard rocks. If synthetic liner materials are used, a binding layer of suitable fine-grained material should be laid to support the liner. However, if the supporting layer consists of low permeable material (e.g., clay), the synthetic liner must be placed on top of this layer. A layer of similar fine-grained material with the thickness of 25–30 cm should also be laid above the liner to protect it from subsequent mechanical and environmental damage. During the early phase of operation, particular care should be taken to ensure that the traffic does not damage the liner. Monitoring the quality of groundwater close to the site is necessary to get the feedback on the performance of a liner.

(iv) **Leachate and landfill gas management:** The basic elements of the leachate collection system (i.e., drain pipes, drainage layers, collection pipes, sumps, etc.) must be installed immediately above the liner, before any waste is deposited. Particular care must also be taken to prevent the drain and collection pipes from settling. During landfill operations, waste cells are covered with soil to avoid additional contact between waste and the environment. The soil layers have to be sufficiently permeable to allow downward leachate transport. Landfill gas is not extracted before completion, which includes construction of final cover, of the waste body. Extraction wells (diameter 0.3 to 1.0 m) may be constructed during or after operation.

(v) **Landfill capping:** Capping is required to control and minimise leachate generation (by minimising water ingress into the landfill) and facilitate landfill gas control or collection (by installing a low permeability cap over the whole site). A cap may consist of natural (e.g., clay) or synthetic (e.g., poly-ethylene) material with thickness of at least 1 m. An uneven settlement of the waste may be a major cause of cap failure. Designs for capping should, therefore, include consideration of leachate and landfill gas collection wells or vents. For the cap to remain effective, it must be protected from agricultural machinery, drying and cracking, plant root penetration, burrowing animals and erosion.

4.7.2 Operation

To secure public acceptability, landfill operations require careful planning and determination of the extent of environmental effects. The basic factor influencing the planning of site operations is the nature and quantity of incoming wastes. The various aspects of this include the following:

(i) **Methods of filling:** The following variations in land filling techniques are available (Burner and Kelly, 1972):

- **Trench method:** This involves the excavation of a trench into which waste is deposited, and the excavated material is then used as cover.
- **Area method:** Wastes may be deposited in layers and so form terraces over the available area. However, with this type of operation, excessive leachate generation may occur, which may render the control difficult.
- **Cell method:** This method involves the deposition of wastes within pre-constructed bounded area. It is now the preferred method in the industrialised world, since it encourages the concept of progressive filling and restoration. Operating a cellular method of filling enables wastes to be deposited in a tidy manner, as the cells serve both to conceal the tipping operation and trap much of the litter that has been generated.
- **Canyon/depression:** This method refers to the placing of suitable wastes against lined canyon or ravine slide slopes. (Slope stability and leachate gas emission control are critical issues for this type of waste placement.)

Figure 4.4 illustrates the land filling methods touched upon above.

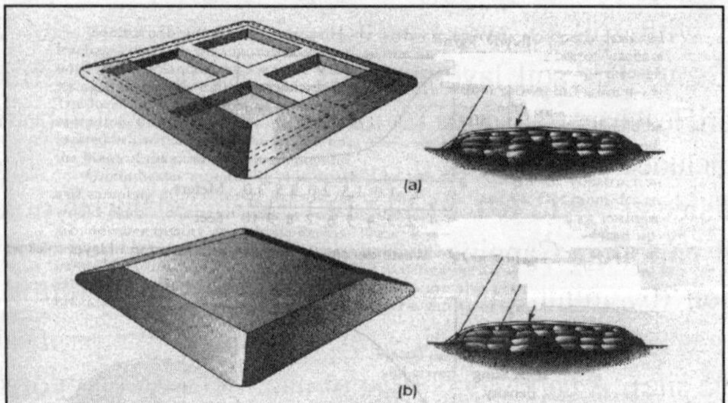

Figure 4.4 Commonly Used Land Filling Methods: (a) trench, (b) area and (c) canyon/depression methods

(ii) **Refuse placement:** The working space should be sufficiently extensive to permit vehicles to manoeuvre and unload quickly and safely without impeding refuse spreading, and allow easy operation of the site equipment. Depositing waste in thin layers and using a compactor enables a high waste density to be achieved. Each progressive layer should not be more than 30 cm thick. The number of passes by a machine over the waste determines the level of compaction.

(iii) **Covering of waste:** At the end of each working day, all exposed surfaces, including the flanks and working space, should be covered with a suitable inert material to a depth of at least 15 cm. This daily cover is considered essential, as it minimises windblown litter and helps reduce odours. Cover material may be obtained from on-site excavations or inert waste materials coming to the site. Pulverised fuel ash or sewage sludge can also be used for this purpose.

(iv) **Site equipment and workforce orientation:** The equipment most commonly used on landfill sites includes steel wheeled compactors, tracked dozers, loaders, earthmovers and hydraulic excavators. Scrapers are used for excavating and moving cover materials. In addition to appropriate equipment, proper training must be ensured for the workforce. They should be competent, and adequately supervised; training should include site safety and first aid. Since a landfill site may pose dangers to both site operators and users, it is necessary to lay down emergency plans and test them from time to time (Phelps et al., 1995).

4.7.3 Monitoring

Landfill represents a complex process of transforming polluting wastes into environmentally acceptable deposits. Because of the complexity of these processes and their potential environmental effects, it is imperative to monitor and confirm that the landfill works as expected. A monitoring scheme, for example, is required for collecting detailed information on the development of leachate and landfill gas within and beyond a landfill. The scheme should be site specific, drawn at the site investigation stage and implemented. Monitoring is generally done for the following:

(i) **Leachate/gas:** Monitoring of leachate/gas plays a vital role in the management of landfills. Data on the volume of leachate/gas and their composition are essential for proper control of leachate/gas generation and its treatment. Knowledge of the chemical composition of leachate/gas is also required to confirm that attenuation processes within the landfill are proceeding as expected. Various systems for monitoring the leachate level are in use, and are mostly based on pipes installed prior to land filling. Note that small bore perforated plastic pipes are relatively cheaper and easier to install, but have the disadvantage of getting damaged faster during infilling. Placing pipes within a column or tyres may, however, offer some protection.

(ii) **Groundwater:** A continued groundwater-monitoring programme for confirming the integrity of the liner system is essential. At an early stage of site preparation, therefore, a number of monitoring boreholes need to be provided around the site. However, the location, design and number of boreholes depend on the size of the landfill, proximity to an aquifer, geology of the site and types of wastes deposited. Installation of a double liner system can make the monitoring exercise more accurate and easier to perform. Water should be regularly flushed through the secondary leachate collection system. In case this water is polluted, the primary leachate barrier will be damaged, and if repair is not considered possible, the leachate collected must be transported to the leachate treatment facility.

✍ LEARNING ACTIVITY 4.7

Explain the need of monitoring sanitary landfills.

Note:

a) Write your answer in the space given below.
b) Check your answer with the one given at the end of this book.

4.8 WASTE DISPOSAL: A CASE STUDY OF BANGALORE

One of the critical concerns of a municipal corporation is planning for a proper waste disposal in response to the increasing volume and hazardous nature of urban wastes. When wastes are disposed off unhygienically, they do spoil the aesthetic value of the city as well as create problems such as breeding of pathogenic organisms, which serve as carriers of diseases (Attarwalla, 1993; Areivala, 1971). Some of the principal problems associated with disposal of solid wastes can be categorised as under:

- Diseases, i.e., rats, flies and other pests feed on the wastes and carry diseases.
- Air/noise pollution, e.g., increase in vehicular traffic, smoke, fly ash and odours.
- Ground and surface water pollution, e.g., runoff during the monsoon season causes surface water pollution, while percolation often causes groundwater contamination.
- Unaesthetic appearance because of litter (Gotoh, 1989).

However, we can minimise or satisfactorily deal with these problems through competent engineering and planning, selecting appropriate waste disposal sites and methods of operation, and making SWM strategies essentially local (see also *http://stratema.sigis.net/cupum/pdf/E1.pdf*).

Against this backdrop, let us now assess the scenario in Bangalore. About two-thirds of the waste (about 1600 tonnes/day) in the Bangalore city is getting dumped in the outskirts of the city. As there are no sanitary landfills in the city for proper dumping of waste, it is merely transported to the outskirts and disposed off in any abandoned open land, usually along public highways (Vagale, 1997). The Bangalore Mahanagara Palike (BMP) along with the Karnataka State Pollution Control Board (KSPCB) has, however, identified nine abandoned quarries around the city for sanitary landfills. Table 4.3 contains the list of these sites:

Table 4.3 Solid Waste Disposal Sites Identified by the BMP

Sl No.	*Name of site*	*Area (acres)*
1	B. Narayanpura	10.15
2	Vibuthipura	8.01
3	Devanachikkanahalli	6.09
4	Sarakki	2.00
5	Hongasandra	4.02
6	Lakkasandra	10.04
7	Hennur	10.00
8	Kudittally	0.36
9	Adugodi	2.00

Source: Department of SWM, Bangalore Mahanagara Palike, Bangalore, 1998

Of the sites listed in Table 4.3, only three have been selected after an assessment of suitability, viz. B. Narayanpura (situated about 10 km northeast of the city in Krishnarajapura hobli), Hennur (situated at a distance of about 9 km north of the city) and Devanachikkanahalli (situated about 10 km to the southeast of the city). These sites were selected on the basis of the geo-technical assessment carried out after a site visit and review of data. However, a periodical assessment of ground water and air quality, before and during the process of land filling, is necessary.

SUMMARY

In this Unit, we discussed some of the problems associated with the indiscriminate disposal of wastes and the key issues involved in safe disposal. In this context, we said that safe disposal is constrained by various issues such as inadequate municipal capacity, lack of political commitment, lack of finance, insufficient technical guidelines and lack of accountability of individuals and institutions. We, then, discussed some of the salient features of waste disposal including various disposal options such as uncontrolled dumping, sanitary landfill, composting, incineration, gasification, refuse-derived fuel and pyrolysis, and their selection criteria (i.e., technical, institutional, financial, social and environmental). Having given a background of waste disposal options, we discussed in detail sanitary landfill in terms of principle, processes, environmental effects, design, construction, operation and monitoring. We closed the Unit by giving a case study of Bangalore.

SUGGESTED READINGS

Tchobanoglous, G., Theisen, H. and Eliassan, R. 1977. Solid Wastes—Engineering Principles and Management Issues, McGraw-Hill Series, Kogakusha, Ltd.

US Environmental Protection Agency. 1989. *Decision-Maker's Guide to Solid Waste Management, Vol I*, Washington.

US Environmental Protection Agency. 1995. *Decision-Maker's Guide to Solid Waste Management, Vol II*, Washington.

Arne Vesilind, William Worrel and Reinhart Debra. 2002. Solid Waste Engineering. Thomson Brooks/Cole, Singapore.

Tchobanoglous, G., Theisen, H. and Samuel, A. Vigil, 1993. Integrated Solid Waste Management. McGraw-Hill, Inc., New Delhi.

http//ces.iisc.ernet.in/energy/SWMTR/TR85.html

http://stratema.sigis.net/cupum/pdf/E1.pdf

REFERENCES

Ali, S.M., Cotton, A.P. and Westlake, K. 1999. Down to Earth: Solid Waste Disposal for Low-Income Countries. WEDC, Loughborough University, UK.

Areivala, S.J. 1971. Solid Wastes Disposal in India. Central Public Health Engineering Institute, Nagpur.

Attarwalla, F. A. 1993. Solid Waste Management - A Manual. All India Institute of Local Self-Government, Andheri (W), Bombay, India.

Burner, D.R. and Kelly, D.J. 1972. Sanitary Landfill Design and Operation. EPA office of Solid Waste Management, Washington D.C.

Gotoh, S. 1989. Solid Waste Management: A Base for Healthier Metropolitan Development, Editorial Introduction, Regional Development Dialogue, An International Journal Focusing on Third World Development Problems, United Nations Centre for Regional Studies 18.

Phelps, H.O., Heinke. G.W., Jonker, J.F., Ouano, E.A.R. and Vandecasteele, C. 1995. Management of Solid Wastes, UNESCO, Paris.

Vagale, L.R. 1997. Environment of Urban Areas in India Case Study: Bangalore. ENVIS Journal of Human Settlements, Centre for Environmental Studies, School of Planning and Architecture, New Delhi, India, November 1997.

http://stratema.sigis.net/cupum/pdf/E1.pdf

UNIT 5

Waste Processing Techniques

STRUCTURE

OVERVIEW

In Unit 4, we discussed conventional and engineered waste disposal options, and also mentioned that through proper processing, we would be able to recover resource and energy from wastes. In Unit 5, we will explain some of the important techniques used for processing solid wastes for the recovery of materials, and their design criteria. The processing techniques we will be discussing in this Unit include mechanical and chemical volume reduction, component separation, and drying and dewatering.

LEARNING OBJECTIVES

After completing this Unit, you should be able to:

- identify the purpose of waste processing;
- explain the processing techniques for reducing the volume and size of wastes;

- carry out separation of various components;
- discuss the need for dewatering and drying of wastes;
- assess technical viability of various processing techniques.

5.1 PURPOSE OF PROCESSING

The processing of wastes helps in achieving the best possible benefit from every functional element of the solid waste management (SWM) system and, therefore, requires proper selection of techniques and equipment for every element. Accordingly, the wastes that are considered suitable for further use need to be paid special attention in terms of processing, in order that we could derive maximum economical value from them.

The purposes of processing, essentially, are (Tchobanoglous et al., 1993):

(i) **Improving efficiency of SWM system:** Various processing techniques are available to improve the efficiency of SWM system. For example, before waste papers are reused, they are usually baled to reduce transporting and storage volume requirements. In some cases, wastes are baled to reduce the haul costs at disposal site, where solid wastes are compacted to use the available land effectively. If solid wastes are to be transported hydraulically and pneumatically, some form of shredding is also required. Shredding is also used to improve the efficiency of the disposal site.

(ii) **Recovering material for reuse:** Usually, materials having a market, when present in wastes in sufficient quantity to justify their separation, are most amenable to recovery and recycling. Materials that can be recovered from solid wastes include paper, cardboard, plastic, glass, ferrous metal, aluminium and other residual metals. (We will discuss some of the recovery techniques later in Section 5.3.)

(iii) **Recovering conversion products and energy:** Combustible organic materials can be converted to intermediate products and ultimately to usable energy. This can be done either through incineration, pyrolysis, composting or bio-digestion. Initially, the combustible organic matter is separated from the other solid waste components. Once separated, further processing like shredding and drying is necessary before the waste material can be used for power generation. (We will explain these energy recovery techniques in Units 7 and 8.)

Having described the need for waste processing, we now discuss how waste processing is actually carried out.

5.2 MECHANICAL VOLUME AND SIZE REDUCTION

Mechanical volume and size reduction is an important factor in the development and operation of any SWM system. The main purpose is to reduce the volume (amount) and size of waste, as compared to its original form, and produce waste of uniform size. We will discuss the processes involved in volume and size reduction along with their selection criteria, equipment requirement, design consideration, etc., in Subsections 5.2.1 and 5.2.2.

5.2.1 Volume Reduction or Compaction

Volume reduction or compaction refers to densifying wastes in order to reduce their volume. Some of the benefits of compaction include:

- reduction in the quantity of materials to be handled at the disposal site;
- improved efficiency of collection and disposal of wastes;
- increased life of landfills;
- economically viable waste management system.

However, note the following disadvantages associated with compaction:

- poor quality of recyclable materials sorted out of compaction vehicle;
- difficulty in segregation or sorting (since the various recyclable materials are mixed and compressed in lumps);
- bio-degradable materials (e.g., leftover food, fruits and vegetables) destroy the value of paper and plastic material.

Equipment Used for Compaction

Based on their mobility, we can categorise the compaction equipment used in volume reduction under either of the following:

(i) **Stationary equipment:** This represents the equipment in which wastes are brought to, and loaded into, either manually or mechanically. In fact, the compaction mechanism used to compress waste in a collection vehicle, is a stationary compactor. According to their application, stationary compactors can be described as light duty (e.g., those used for residential areas), commercial or light industrial, heavy industrial and transfer station compactors. Usually, large stationary compactors are necessary, when wastes are to be compressed into:

- steel containers that can be subsequently moved manually or mechanically;

- chambers where the compressed blocks are binded or tied by some means before being removed;
- chambers where they are compressed into a block and then released and hauled away untied;
- transport vehicles directly.

(ii) **Movable equipment:** This represents the wheeled and tracked equipment used to place and compact solid wastes, as in a sanitary landfill.

Table 5.1 below lists the types of commonly used compaction equipment and their suitability.

Table 5.1 Types of Compaction Equipment

Location or Operation	*Type of Compactor Stationary/Residential*	*Remarks*
Solid waste generation points	Vertical	Vertical compaction ram may be used; may be mechanically or hydraulically operated, usually hand-fed; wastes compacted into corrugated box containers, or paper or plastic bags; used in medium and high-rise apartments.
	Rotary	Ram mechanism used to compact waste into paper or plastic bags on rotating platform, platform rotates as containers are filled; used in medium and high-rise apartments.
	Bag or extruder	Compactor can be chute fed; either vertical or horizontal rams; single or continuous multi-bags; single bag must be replaced and continuous bags must be tied off and replaced; used in medium and high-rise apartments.
	Under counter	Small compactors used in individual residences and apartment units; wastes compacted into special paper bags; after wastes are dropped through a panel door into a bag and door is closed, they are sprayed for odour control; button is pushed to activate compaction mechanism.

(*Contd.*)

Table 5.1 (*Contd.*)

Location or Operation	*Type of Compactor Stationary/Residential*	*Remarks*
	Stationary/commercial	Compactor with vertical and horizontal ram; wastes compressed into steel containers; compressed wastes are manually tied and removed; used in low, medium and high-rise apartments, commercial and industrial facilities.
Collection	Stationary/packers	Collection vehicles equipped with compaction mechanism.
Transfer and/or processing station	Stationary/transfer trailer	Transfer trailer, usually enclosed, equipped with self-contained compaction mechanism.
	Stationary low pressure	Wastes are compacted into large containers.
	Stationary high pressure	Wastes are compacted into dense bales or other forms.
Disposal site	Movable wheeled or tracted equipment	Specially designed equipment to achieve maximum compaction of wastes.
	Stationary/track mounted	High-pressure movable stationary compactors used for volume reduction at a disposal site.

Source: Tchobanoglous et al. (1993)

Let us now move on to the discussion of compactors used in the transfer station.

Compactors

According to their compaction pressure, we can divide the compactors used at transfer stations as follows:

(i) **Low-pressure (less than 7 kg/cm^2) compaction:** This includes those used at apartments and commercial establishments, baling equipment used for waste papers and cardboards and stationary compactors used at transfer stations. In low-pressure compaction, wastes are compacted in large containers. Note that portable stationary compactors are being used increasingly by a number of industries in conjunction with material recovery options, especially for waste paper and cardboard.

(ii) **High-pressure (more than 7 kg/cm^2) compaction:** Compact systems with a capacity up to 351.5 kg/cm^2 or 5000 lb/in^2 come under this category. In such systems, specialised compaction equipment are used to compress solid wastes into blocks or bales of various sizes. In some cases, pulverised wastes are extruded after compaction in the form of logs. The volume reduction achieved with these high-pressure compaction systems varies with the characteristics of the waste. Typically, the reduction ranges from about 3 to 1 through 8 to 1.

When wastes are compressed, their volume is reduced, which is normally expressed in percentage and computed by equation 5.1, given below:

$$\text{Volume reduction}\,(\%) = \frac{V_i - V_f}{V_i} \times 100 \qquad (5.1)$$

The compaction ratio of the waste is given in equation 5.2:

$$\text{Compaction ratio} = \frac{V_i}{V_f} \qquad (5.2)$$

where V_i = volume of waste before compaction, m^3 and V_f = volume of waste after compaction, m^3.

The relationship between the compaction ratio and percent of volume reduction is important in making a trade-off analysis between compaction ratio and cost. Other factors that must be considered are final density of waste after compaction and moisture content. The moisture content that varies with location is another variable that has a major effect on the degree of compaction achieved. In some stationary compactors, provision is made to add moisture, usually in the form of water, during the compaction process.

Selection of Compaction Equipment

To ensure effective processing, we need to consider the following factors, while selecting compaction equipment.

- Characteristics such as size, composition, moisture content, and bulk density of the waste to be compacted.
- Method of transferring and feeding wastes to the compactor, and handling.
- Potential uses of compacted waste materials.
- Design characteristics such as the size of loading chamber, compaction pressure, compaction ratio, etc.
- Operational characteristics such as energy requirements, routine and specialised maintenance requirement, simplicity of operation, reliability, noise output, and air and water pollution control requirement.
- Site consideration, including space and height, access, noise and related environmental limitations.

5.2.2 Size Reduction or Shredding

This is required to convert large sized wastes (as they are collected) into smaller pieces. Size reduction helps in obtaining the final product in a reasonably uniform and considerably reduced size in comparison to the original form. But note that size reduction does not necessarily imply volume reduction, and this must be factored into the design and operation of SWM systems as well as in the recovery of materials for reuse and conversion to energy.

In the overall process of waste treatment and disposal, size reduction is implemented ahead of:

- land filling to provide a more homogeneous product. This may require less cover material and less frequent covering than that without shredding. This can be of economic importance, where cover material is scarce or needs to be brought to the landfill site from some distance.
- recovering materials from the waste stream for recycling.
- baling the wastes – a process sometimes used ahead of long distance transport of solid wastes – to achieve a greater density.
- making the waste a better fuel for incineration waste energy recovery facilities. (The size reduction techniques, coupled with separation techniques such as screening, result in a more homogeneous mixture of relatively uniform size, moisture content and heating value, and thereby improving the steps of incineration and energy recovery. We will discuss incineration in Unit 8.)
- reducing moisture, i.e., drying and dewatering of wastes (see Section 5.4 for a discussion on drying and dewatering).

Equipment Used for Size Reduction

Table 5.2 lists the various equipment used for size reduction.

Table 5.2 Size Reduction Equipment

Type	*Mode of action*	*Application*
Small grinders	Grinding, mashing	Organic residential solid wastes
Chippers	Cutting, slicing	Paper, cardboard, tree trimmings, yard waste, wood, plastics
Large grinders	Grinding, mashing	Brittle and friable materials, used mostly in industrial operation
Jaw crushers	Crushing, breaking	Large solids

(*Contd.*)

Table 5.2 (*Contd.*)

Rasp mills	Shredding, tearing	Moistened solid wastes
Shredders	Shearing, tearing	All types of municipal wastes
Cutters, Clippers	Shearing, tearing	All types of municipal wastes
Hammer mills	Breaking, tearing, cutting, crushing	All types of municipal wastes, most commonly used equipment for reducing size and homogenizing composition of wastes
Hydropulper	Shearing, tearing	Ideally suited for use with pulpable wastes, including paper, wood chips. Used primarily in the papermaking industry. Also used to destroy paper records

The most frequently used shredding equipment are the following:

(i) **Hammer mill:** These are used most often in large commercial operations for reducing the size of wastes. Hammer mill is an impact device consisting of a number of hammers, fastened flexibly to an inner disk, as shown in Fig. 5.1, which rotates at a very high speed.

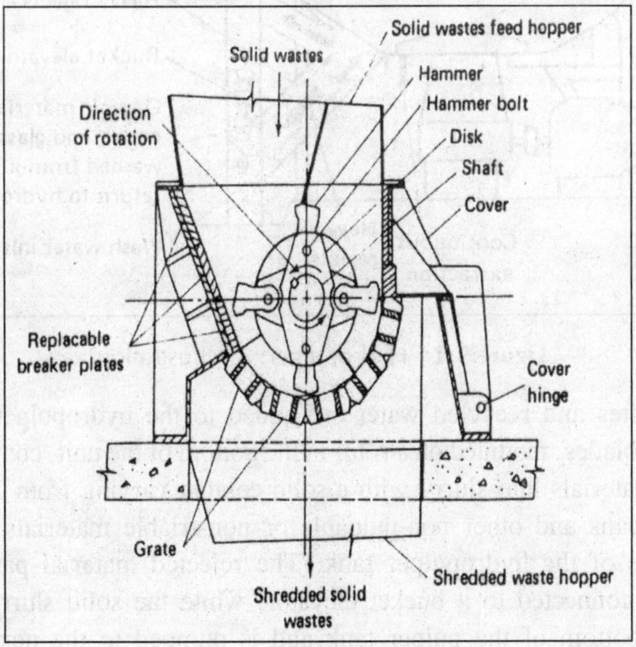

Figure 5.1 Hammer Mill: An Illustration

Solid wastes, as they enter the mill (see Fig. 5.1), are hit by sufficient force, which crush or tear them with a velocity so that they do not adhere to the hammers. Wastes are further reduced in size by being struck between breaker plates and/or cutting bars fixed around the periphery of the inner chamber. This process of cutting and striking action continues, until the required size of material is achieved and after that it falls out of the bottom of the mill.

(ii) **Hydropulper:** An alternative method of size reduction involves the use of a hydropulper as shown in Fig. 5.2.

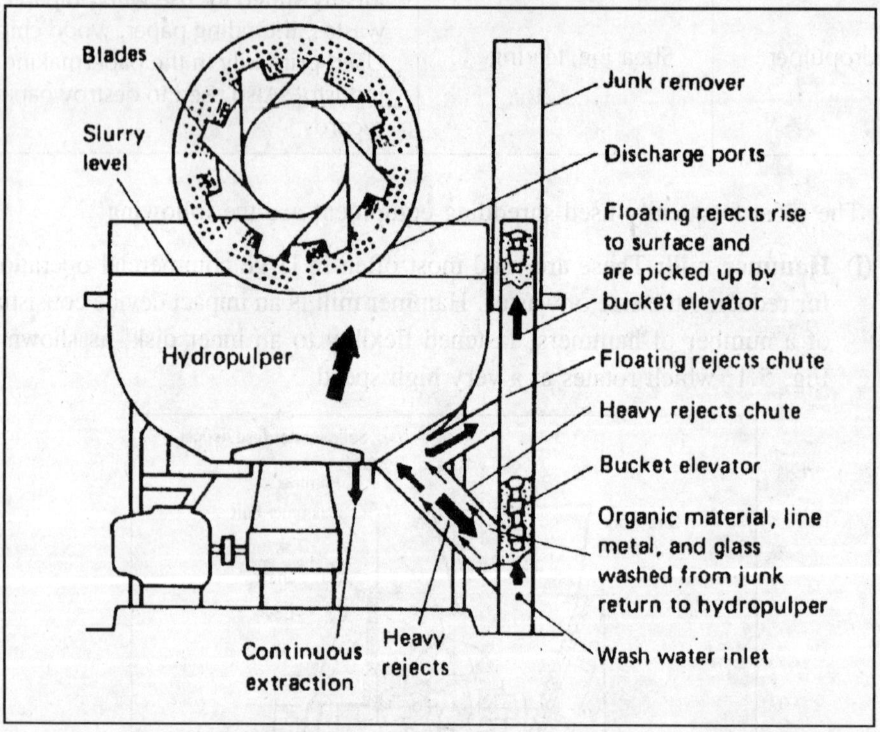

Figure 5.2 Hydropulper: An Illustration

Solid wastes and recycled water are added to the hydropulper. The high-speed cutting blades, mounted on a rotor in the bottom of the unit, convert pulpable and friable materials into slurry with a solid content varying from 2.5 to 3.5%. Metal, tins, cans and other non-pulpable or non-friable materials are rejected from the side of the hydropulper tank. The rejected material passes down a chute that is connected to a bucket elevator, while the solid slurry passes out through the bottom of the pulper tank and is pumped to the next processing operation.

Selection of Size Reduction Equipment

The factors that decide the selection of size reduction equipment include the following.

- The properties of materials before and after shredding.
- Size requirements for shredded material by component.
- Method of feeding shredders, provision of adequate shredder hood capacity (to avoid bridging) and clearance requirement between feed and transfer conveyors and shredders.
- Types of operation (continuous or intermittent).
- Operational characteristics including energy requirements, routine and specialised maintenance requirement, simplicity of operation, reliability, noise output, and air and water pollution control requirements.
- Site considerations, including space and height, access, noise and environmental limitations.
- Metal storage after size reduction for the next operation.

✍ LEARNING ACTIVITY 5.1

Explain the difference between compaction and size reduction and their importance in SWM.

Note:

a) Write your answer in the space given below.

b) Check your answer with the one given at the end of this book.

Besides mechanical techniques of compaction and shredding to reduce the volume and size of wastes, there are also chemical processes through which we can reduce the volume of wastes, which we will touch upon next.

Chemical Volume Reduction

Chemical volume reduction is a method, wherein volume reduction occurs through chemical changes brought within the waste either through an addition of chemicals or changes in temperature. Incineration is the most common method used to reduce the volume of waste chemically, and is used both for volume reduction and power production. The other chemical methods used to reduce volume of waste chemically include *pyrolysis*, *hydrolysis* and chemical conversions. (We will discuss incineration and related issues in Unit 8.)

Note that prior to size or volume reduction (we discussed in Section 5.2), component separation is necessary to avoid the problem of segregating or sorting recyclable materials from the mixed and compressed lumps of wastes and the poor quality of recyclable materials sorted out of compaction vehicles. We will discuss component separation in Section 5.3.

5.3 COMPONENT SEPARATION

Component separation is a necessary operation in which the waste components are identified and sorted either manually or mechanically to aid further processing. This is required for the:

- recovery of valuable materials for recycling;
- preparation of solid wastes by removing certain components prior to incineration, energy recovery, composting and biogas production. (Note that these are discussed in Units 8 and 9.)

The most effective way of separation is manual sorting in households prior to collection. In many cities (e.g., Bangalore, Chennai, etc., in India), such systems are now routinely used. The municipality generally provides separate, easily identifiable containers into which the householder deposits segregated recyclable materials such as paper, glass, metals, etc. Usually, separate collections are carried out for the recyclable material. At curbside, separate areas are set aside for each of the recyclable materials for householders to deliver material – when there is no municipal collection system. In case the separation is not done prior to collection, it could be sorted out through mechanical techniques such as air separation, magnetic separation, etc., to recover the wastes. We will discuss some of these techniques in Subsections 5.3.1 to 5.3.4.

5.3.1 Air Separation

This technique has been in use for a number of years in industrial operations for segregating various components from dry mixture. Air separation is primarily used to separate lighter materials (usually organic) from heavier (usually inorganic) ones. The lighter material may include plastics, paper and paper products and other organic materials. Generally, there is also a need to separate the light fraction of organic material from the conveying air streams, which is usually done in a cyclone separator. In this technique, the heavy fraction is removed from the air classifier (i.e., equipment used for air separation) to the recycling stage or to land disposal, as appropriate. The light fraction may be used, with or without further size reduction, as fuel for incinerators or as compost material. There are various types of air classifiers commonly used, some of which are listed below.

(i) **Conventional chute type:** This, as shown in Fig. 5.3, is one of the simplest types of air classifiers.

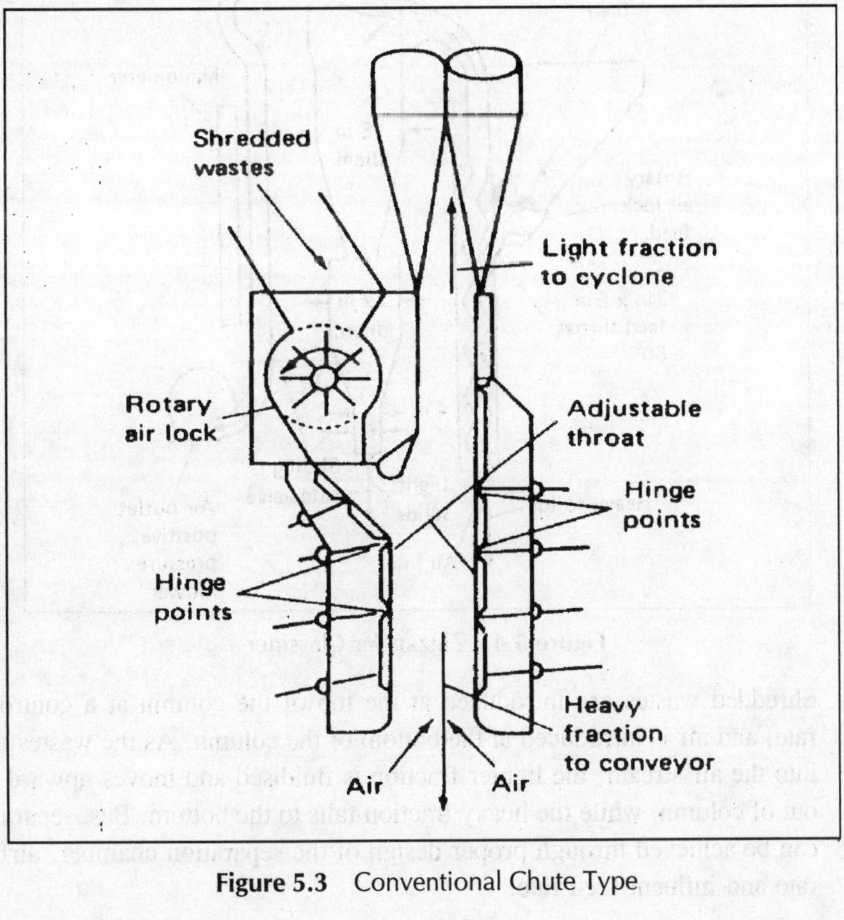

Figure 5.3 Conventional Chute Type

In this type, when the processed solid wastes are dropped into the vertical chute, the lighter material is carried by the airflow to the top while the heavier materials fall to the bottom of the chute. The control of the percentage split between the light and heavy fraction is accomplished by varying the waste loading rate, airflow rate and the cross section of chute. A rotary air lock feed mechanism is required to introduce the shredded wastes into the classifier.

(ii) **Zigzag air classifier:** An experimental zigzag air classifier, shown in Fig. 5.4 below, consists of a continuous vertical column with internal zigzag deflectors through which air is drawn at a high rate.

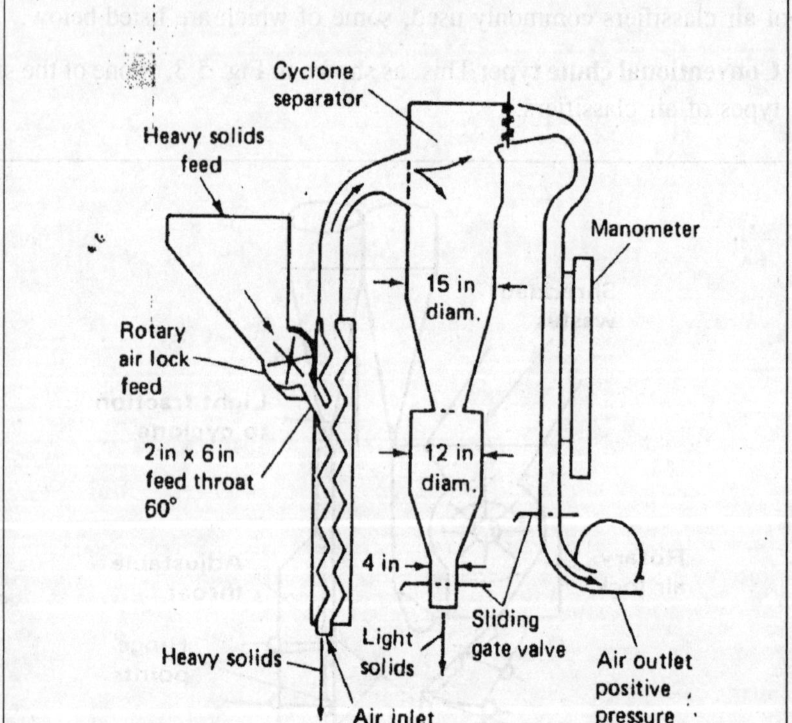

Figure 5.4 Zigzag Air Classifier

Shredded wastes are introduced at the top of the column at a controlled rate, and air is introduced at the bottom of the column. As the wastes drop into the air stream, the lighter fraction is fluidised and moves upward and out of column, while the heavy fraction falls to the bottom. Best separation can be achieved through proper design of the separation chamber, airflow rate and influent feed rate.

(iii) **Open inlet vibrator type:** Figure 5.5 below illustrates this type of air classifier.

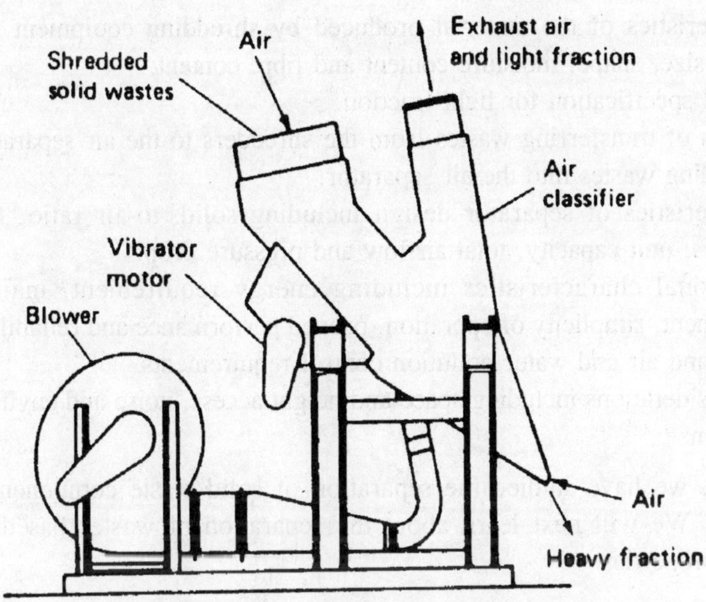

Figure 5.5 Open Inlet Vibrator

In this type of air classifier, the separation is accomplished by a combination of the following actions.

- **Vibration:** This helps to stratify the material fed to the separator into heavy and light components. Due to this agitation, the heavier particles tend to settle at the bottom as the shredded waste is conveyed down the length of the separator.
- **Inertial force:** In this action, the air pulled in through the feed inlet imparts an initial acceleration to the lighter particle, while the wastes travel down the separator as they are being agitated.
- **Air pressure:** This action refers to the injection of fluidising air in two or more high velocity and low mass flow curtains across the bed. A final stripping of light particles is accomplished at the point where the heavy fraction discharges from the elutriators. It has been reported that the resulting separation is less sensitive to particle size than a conventional vertical air classifier, be it of straight or zigzag design. An advantage of this classifier is that an air lock feed mechanism is not required and wastes are fed by gravity directly into the separator inlet.

Selection of Air Separation Equipment

The factors that are to be considered for selecting air separation equipment include the following:

- Characteristics of the material produced by shredding equipment including particle size, shape, moisture content and fibre content.
- Material specification for light fraction.
- Methods of transferring wastes from the shredders to the air separation units and feeding wastes into the air separator.
- Characteristics of separator design including solids-to-air ratio, fluidising velocities, unit capacity, total airflow and pressure drop.
- Operational characteristics including energy requirement, maintenance requirement, simplicity of operation, proved performance and reliability, noise output, and air and water pollution control requirements.
- Site considerations including space and height access, noise and environmental limitations.

So far, we have studied the separation of solid waste components by air separation. We will next learn about the separation of wastes based on their magnetic properties.

5.3.2 Magnetic Separation

The most common method of recovering ferrous scrap from shredded solid wastes involves the use of magnetic recovery systems. Ferrous materials are usually recovered either after shredding or before air classification. When wastes are mass-fired in incinerators, the magnetic separator is used to remove the ferrous material from the incinerator residue. Magnetic recovery systems have also been used at landfill disposal sites. The specific locations, where ferrous materials are recovered will depend on the objectives to be achieved, such as reduction of wear and tear on processing and separation equipment, degree of product purity achieved and the required recovery efficiency.

Equipment Used for Magnetic Separation

Various types of equipment are in use for the magnetic separation of ferrous materials. The most common types are the following:

(i) **Suspended magnet:** In this type of separator, a permanent magnet is used to attract the ferrous metal from the waste stream. When the attracted metal reaches the area, where there is no magnetism, it falls away freely. This ferrous metal is then collected in a container. Figure 5.6 shows a typical suspended magnet.

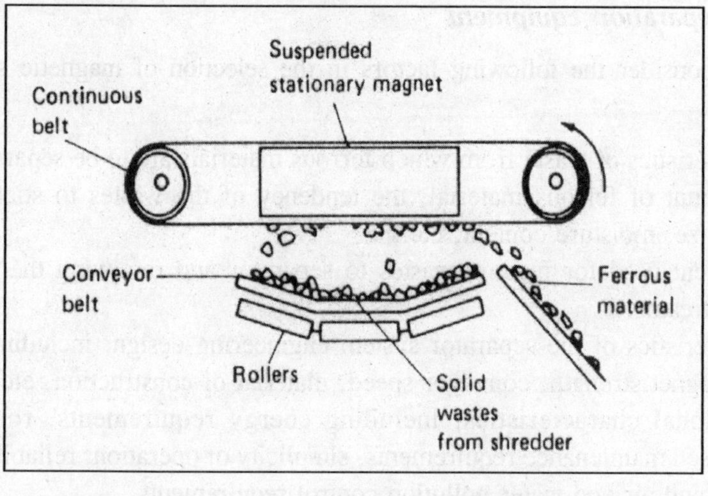

Figure 5.6 Suspended Type Permanent Magnetic Separator

This type of separation device is suitable for processing raw refuse, where separators can remove large pieces of ferrous metal easily from the waste stream.

(ii) **Magnetic pulley:** This consists of a drum type device containing permanent magnets or electromagnets over which a conveyor or a similar transfer mechanism carries the waste stream. The conveyor belt conforms to the rounded shape of the magnetic drum and the magnetic force pulls the ferrous material away from the falling stream of solid waste. Figure 5.7 illustrates this type of magnetic separator.

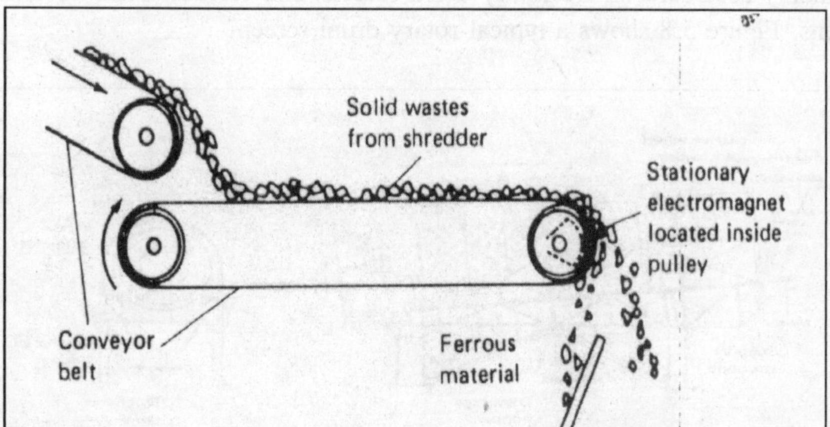

Figure 5.7 Pulley Type Permanent Magnetic Separator

Selection of Magnetic Separation Equipment

We must consider the following factors in the selection of magnetic separation equipment:

- Characteristics of waste from which ferrous materials are to be separated (i.e., the amount of ferrous material, the tendency of the wastes to stick to each other, size, moisture content, etc.)
- Equipment used for feeding wastes to separator and removing the separated waste streams.
- Characteristics of the separator system engineering design, including loading rate, magnet strength, conveyor speed, material of construction, etc.
- Operational characteristics, including energy requirements, routine and specialised maintenance requirements, simplicity of operation, reliability, noise output, and air and water pollution control requirements.
- Locations where ferrous materials are to be recovered from solid wastes.
- Site consideration, including space and height, access, noise and environmental limitations.

5.3.3 Screening

Screening is the most common form of separating solid wastes, depending on their size by the use of one or more screening surfaces. Screening has a number of applications in solid waste resource and energy recovery systems. Screens can be used before or after shredding and after air separation of wastes in various applications dealing with both light and heavy fraction materials. The most commonly used screens are rotary drum screens and various forms of vibrating screens. Figure 5.8 shows a typical rotary drum screen.

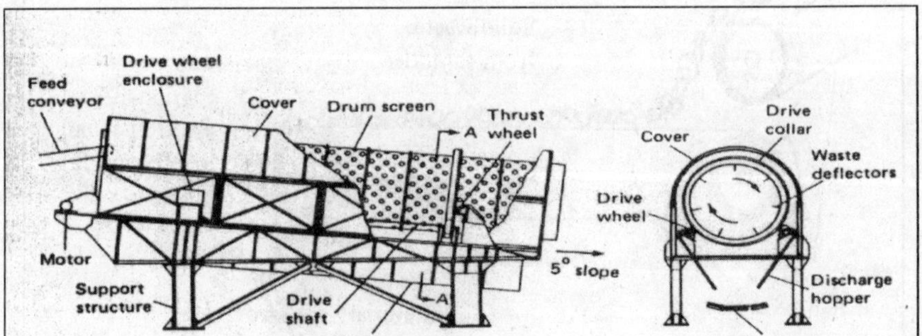

Figure 5.8 Rotary Drum Screen
Source: Tchobanoglous et al. (1977)

Note that rotating wire screens with relatively large openings are used for separation of cardboard and paper products, while vibrating screens and rotating drum screens are typically used for the removal of glass and related materials from the shredded solid wastes.

Selection of Screening Equipment

The various factors that affect the selection of screens include the following:

- Material specification for screened component.
- Location where screening is to be applied and characteristics of waste material to be screened, including particle size, shape, bulk, density and moisture content.
- Separation and overall efficiency.
- Characteristics screen design, including materials of construction, size of screen openings, total surface screening area, oscillating rate for vibrating screens, speed for rotary drum screens, loading rates and length.
- Operational characteristics, including energy requirements, maintenance requirements, simplicity of operation, reliability, noise output and air and water pollution control requirements.
- Site considerations such as space and height access, noise and related environmental limitations.

The efficiency of screen can be evaluated in terms of the percentage recovery of the material in the feed stream by using equation 5.3.

$$\text{Recovery (\%)} = \frac{U \times W_u}{F \times W_f} \times 100 \tag{5.3}$$

$$W_f = \frac{\text{Weight of sample}}{\text{Weight of material fed to the screen}} \tag{5.4}$$

$$W_u = \frac{\text{Weight of sample in underflow}}{\text{Total weight of material in underflow}} \tag{5.5}$$

where U = weight of material passing through screen (underflow), kg/h; F = weight of material fed to the screen, kg/h; W_u = weight fraction of material desired size in underflow and W_f = weight fraction of material of desired size in feed.

The effectiveness of the screening operation can be determined by:

$$\text{Effectiveness} = \text{recovery} \times \text{rejection}$$

where, rejection = 1 – recovery of undesired material

$$= 1 - \frac{U(1 - W_u)}{F(1 - W_f)}$$

Therefore, the effectiveness of screen is:

$$\text{Effectiveness} = \frac{U \times W_u)}{F \times W_f)} \times \left[1 - \frac{U(1 - W_u)}{F(1 - W_f)}\right]$$

LEARNING ACTIVITY 5.2

Given that 100 tonnes/h of solid waste is applied to a rotary screen for the removal of glass prior to shredding, determine the recovery efficiency and effectiveness of the screen, based on the following experimental data:

The percentage of glass in solid waste = 8 %
Total weight of material in underflow = 10 tonnes/h
Weight of glass in screen underflow = 7.2 tonnes/h

Note:

a) Write your answer in the space given below.
b) Check your answer with the one given at the end of this book.

5.3.4 Other Separation Techniques

Besides the mechanical techniques we studied earlier for segregating wastes, there are others. A description of some of the other separation techniques is given below:

(i) **Hand-sorting or previewing:** Previewing of the waste stream and manual removal of large sized materials is necessary, prior to most types of separation or size reduction techniques. This is done to prevent damage or stoppage of equipment such as shredders or screens, due to items such as rugs, pillows, mattresses, large metallic or plastic objects, wood or other construction materials, paint cans, etc.

(ii) **Inertial separation:** Inertial methods rely on ballistic or gravity separation principles to separate shredded solid wastes into light (i.e., organic) and heavy (i.e., inorganic) particles. Figures 5.9 and 5.10 illustrate the modes of operation of two different types of inertial separators.

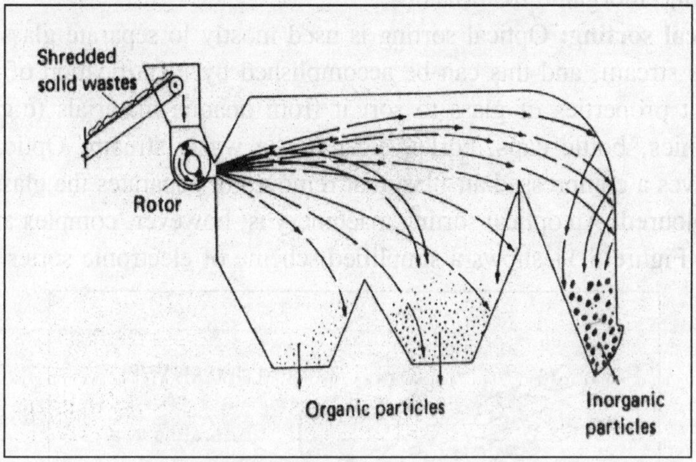

Figure 5.9 Ballistic Inertial Separator

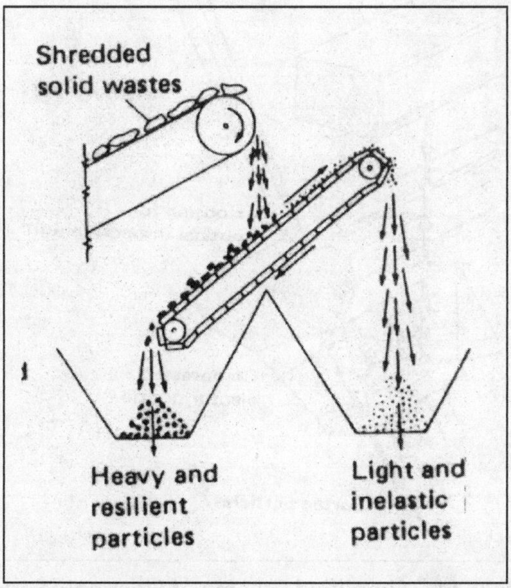

Figure 5.10 Inclined Conveyor Separator

(iii) **Flotation:** In the flotation process, glass-rich feedstock, which is produced by screening the heavy fraction of the air-classified wastes after ferrous metal separation, is immersed in water in a soluble tank. Glass chips, rocks, bricks, bones and dense plastic materials that sink to the bottom are removed with belt scrappers for further processing. Light organic and other materials that float are skimmed from the surface. These materials are taken to landfill sites or to incinerators for energy recovery. Chemical adhesives (flocculants) are also used to improve the capture of light organic and fine inorganic materials.

(iv) **Optical sorting:** Optical sorting is used mostly to separate glass from the waste stream, and this can be accomplished by identification of the transparent properties of glass to sort it from opaque materials (e.g., stones, ceramics, bottle caps, corks, etc.) in the waste stream. Optical sorting involves a compressed air blast that removes or separates the glasses– plain or coloured. An optical sorting machinery is, however, complex and expensive. Figure 5.11 shows a simplified scheme of electronic sorter for glass.

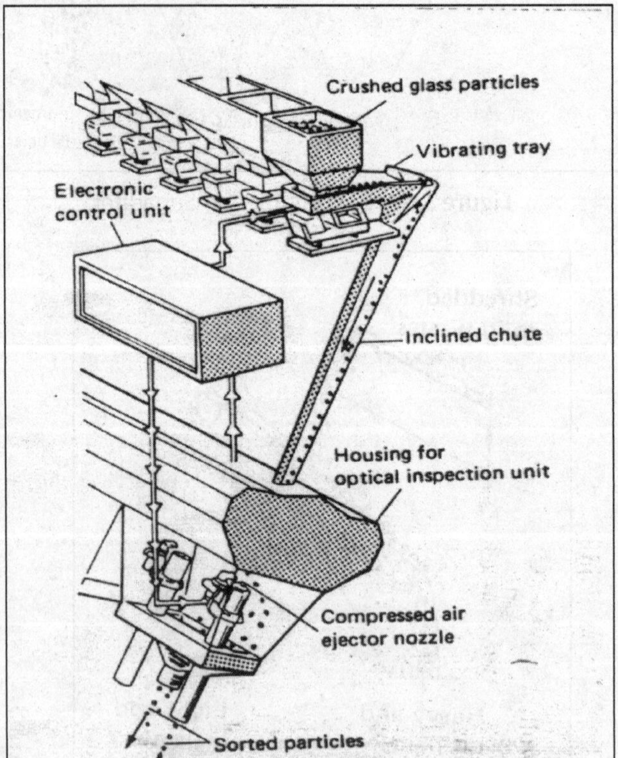

Figure 5.11 Simplified Scheme of Electronic Sorter
Source: Tchobanoglous et al. (1993)

So far, we discussed component separation through air classifiers, magnetic separators, screens, and hand sorting, flotation, optical sorting and inertial separators. In case, however, the waste consists of moisture, we need to remove it for efficient management. It is in this regard that drying and dewatering are considered the most appropriate means of removal of moisture. We will study this next.

5.4 DRYING AND DEWATERING

Drying and dewatering operations are used primarily for incineration systems, with or without energy recovery systems. These are also used for drying of sludges in wastewater treatment plants, prior to their incineration or transport to land disposal. The purpose of drying and dewatering operation is to remove moisture from wastes and thereby make it a better fuel. Sometimes, the light fraction is *pelletised* after drying to make the fuel easier to transport and store, prior to use in an incinerator or energy recovery facility.

Table 5.3 shows the range of moisture content for municipal solid waste components.

Table 5.3 Moisture Content of Municipal Solid Waste Components

Component	*Moisture (in percent)*	
	Range	*Typical*
Food wastes	50–80	70
Paper	4–10	6
Cardboard	4–8	5
Plastics	1–4	2
Textiles	6–15	10
Rubber	1–4	2
Leather	8–12	10
Garden trimmings	30–80	60
Wood	15–40	20
Glass	1–4	2
Tin cans	2–4	3
Nonferrous metals	2–4	2
Ferrous metals	2–6	3
Dirt, ashes, brick, etc.	6–12	8
Municipal solid wastes	15–40	20

Source: Tchobanoglous et al. (1993)

5.4.1 Drying

The following three methods are used to apply the heat required for drying the wastes:

(i) **Convection drying:** In this method, hot air is in direct contact with the wet solid waste stream.

(ii) **Conduction drying:** In this method, the wet solid waste stream is in contact with a heated surface.

(iii) **Radiation drying:** In this method, heat is transmitted directly to the wet solid waste stream by radiation from the heated body.

Of these three methods, convection drying is used most commonly. Figure 5.12 illustrates a rotary drum dryer used in the cement industry.

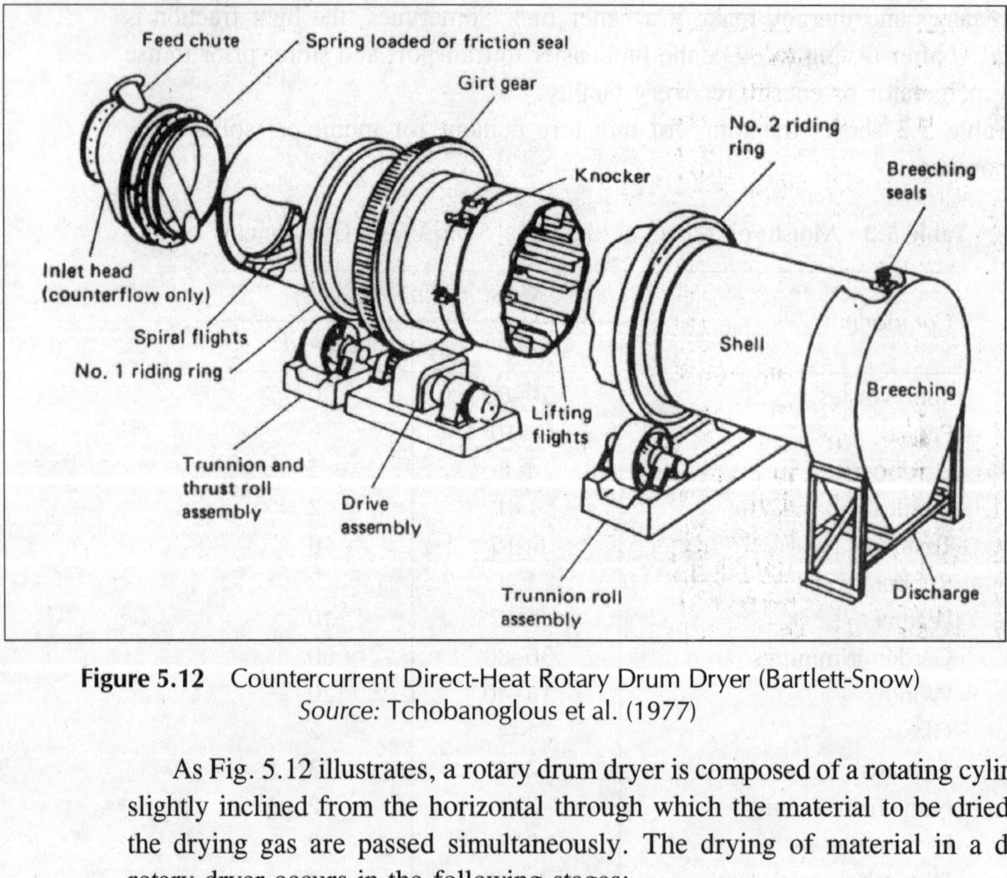

Figure 5.12 Countercurrent Direct-Heat Rotary Drum Dryer (Bartlett-Snow)
Source: Tchobanoglous et al. (1977)

As Fig. 5.12 illustrates, a rotary drum dryer is composed of a rotating cylinder, slightly inclined from the horizontal through which the material to be dried and the drying gas are passed simultaneously. The drying of material in a direct rotary dryer occurs in the following stages:

- Heating the wet material and its moisture content to the constant-rate drying temperature.

- Drying the material substantially at this temperature.
- Heating of material to its discharge temperature and evaporation of moisture remaining at the end of the stage.

The retention time in the rotary drum is about 30 – 45 minutes. The required energy input will depend on the moisture content, and the required energy input can be estimated by using a value of about 715 KJ/kg (or 1850 Btu/lb) of water evaporated. Some of the factors we need to consider in the selection of a drying equipment include the following:

- Properties of material to be dried.
- Drying characteristics of the materials, including moisture content, maximum material temperature and anticipated drying time.
- Specification of final product, including moisture content.
- Nature of operation, whether continuous or intermittent.
- Operational characteristics, including energy requirements, maintenance requirements, simplicity of operation, reliability, noise output and air and water pollution control requirements.
- Site considerations such as space and height access, noise and environmental limitations.

5.4.2 Dewatering

Dewatering is more applicable to the problem of sludge disposal from wastewater treatment of plants, but may also be applicable in some cases to municipal/industrial waste problems. When drying beds, lagoons or spreading on land are not feasible, other mechanical means of dewatering are used. The emphasis in the dewatering operation is often on reducing the liquid volume. Once dewatered, the sludge can be mixed with other solid waste, and the resulting mixture can be:

- incinerated to reduce volume;
- used for the production of recoverable by-products;
- used for production of compost;
- buried in a landfill.

Centrifugation and filtration are the two common methods for the dewatering of sludge. Sludges with solid content of a few percent can be thickened to about 10–15% in centrifugation and about 20–30% in pressure filtration or vacuum filtration.

LEARNING ACTIVITY 5.3

List the methods of drying.

Note:

a) Write your answer in the space given below.

b) Check your answer with the one given at the end of this book.

SUMMARY

In this Unit, we discussed various processing techniques that are used in SWM system to improve the efficiency of operation, recovery of resources, i.e., usable materials, and recovery of conversion product and energy. We began our discussion with the importance of processing techniques and the nature of equipment involved for the purpose. Subsequently, we discussed mechanical volume and size reduction techniques and touched upon chemical volume reduction. We also explained some component separation techniques (air separation, magnetic separation, screening, etc.). We closed the Unit with a discussion on drying and dewatering, i.e., the processing techniques used for removing varying amounts of moisture present in solid wastes.

SUGGESTED READINGS

Areivala, S.J. 1971. *Solid Wastes Disposal in India*, Central Public Health Engineering Institute, Nagpur.

Environmental Management in Developing Countries. 1995. *Waste Management Vol II*, Institute for Scientific Co-operation, Tubingen.

Flintoff, R. 1984. *Management of Solid Wastes in Developing countries*, WHO Regional Publications, Southeast Asia Series No 1.

Arne Vesilind, William Worrel and Reinhart Debra. 2002. Solid Waste Engineering, Thomson Brooks/Cole, Singapore.

Ramachandra, T.V. and Saira Varghese, K. 2004. Environmentally sound options for e-wastes management, ENVIS Journal of Human Settlements (ISSN 0971-9091), March 2004.

Ramachandra, T.V. and Saira Varghese, K. 2003. Exploring possibilities of achieving sustainability in solid waste management, *Indian J. Environmental Health*, **45 (4):** 255-264.

Tchobanoglous, G., Theisen, H. and Samuel, A. Vigil. 1993. Integrated Solid Waste Management, McGraw-Hill, Inc., New Delhi.

UNEP. 1996. *International Source Book on Environmentally Sound Technologies for Municipal Solid Waste Management (6),* IETC, Osaka/Shiga.

US Environmental Protection Agency. 1989. *Decision-Maker's Guide to Solld Waste Management, Vol I*, Washington.

US Environmental Protection Agency.1995. *Decision-Maker's Guide to Solid Waste Management, Vol II*, Washington.

http://ces.iisc.ernet.in/energy/SWMTR/TR85.html

Central Public Health and Environmental Engineering Organisation (CPHEEO) and Ministry of Urban Development, Government of India, New Delhi (2000). Manual on Municipal Solid Waste Management

REFERENCES

Phelps, H.O., Heinke, G. W., Jonker, J.F., Ouano, E.A.R. and Vandecasteele, C. 1995. *Management of Solid Wastes*, UNESCO, Paris.

Tchobanoglous, G., Theisen, H. and Eliassan, R. 1977. *Solid Wastes—Engineering Principles and Management Issues*, McGraw-Hill Book Company, New York.

Tchobanoglous, G., Theisen, H. and Samuel, A. Vigil. 1993. *Integrated Solid Waste Management*, McGraw-Hill, Inc., New Delhi.

UNIT 6

Source Reduction, Product Recovery and Recycling

STRUCTURE

OVERVIEW

As you recall, the focus of our discussion in Unit 5 was on some of the processing techniques in use in SWM systems to improve the efficiency of operations and recover resources (that could be reused) and energy. Now, in Unit 6, we analyse the recovery and recycling processes. We will, accordingly, begin the Unit by discussing the various aspects of source reduction including the design and implementation of specific source reduction programmes that inform decision-making. We will then examine how product recovery and recycling play a critical role in waste management plans. In this context, we will discuss the factors (e.g., material markets, building local expertise and fostering public participation) that are needed for planning an effective recycling programme and the elements of the recycling programme such as source separation, collection, material recovery facilities and processing. We will also look into the processing of some of the common materials that are recycled. We will close the unit by presenting a case study on source reduction, resource recovery and recycling carried out in Bangalore.

LEARNING OBJECTIVES

After completing this unit, you should be able to:

- explain waste reduction;
- discuss the significance of waste recycling;
- plan a recycling programme;
- discuss recycling programme elements;
- identify commonly recycled materials, their use and economic values;
- assess technical options and processes involved in recycling of wastes;
- explain the constraints involved in the implementation of reducing and recycling programmes.

6.1 SOURCE REDUCTION: BASICS

Source reduction, also known as waste prevention, is an approach that precedes waste management and addresses how products are manufactured and purchased. Put differently, this refers to the activities that reduce the amount of waste generated at source as well as activities that involve any change in the design, manufacture, purchase or usage of materials/products to reduce their volume and/or toxicity, before they become part of the solid waste stream (EPA, 1989 and 1995). Reducing waste before it is generated is a logical way to save costs and natural resources, and preserve the local environment. For instance, waste reduction cuts the municipal and commercial costs involved in waste collection and disposal, and improves the productivity by targeting wasteful processes and products.

However, a successful implementation of source reduction programme requires the co-operation of all stakeholders (e.g., businesses, industries, consumers and state and local governments) as the goals and actions of the local waste management system are specific to local conditions. In fact, source reduction programme should be part of a community waste management plan. Source reduction activities vary widely and many factors have to be considered while evaluating them. In Subsections 6.1.1 to 6.1.4, we will examine some of the basics of source reduction processes (EPA, 1989 and 1995).

6.1.1 Purpose

Source reduction can serve several purposes, including the following:

(i) **Product reuse:** Using reusable products, instead of their disposal equivalents, reduce the amount of materials that are to be managed as wastes. An example of product reuse is the reusable shopping bag.

(ii) **Material volume reduction:** Reducing the volume of material used changes the amount of waste entering the waste stream. This helps in controlling the waste generated and its disposal. For example, buying in bulk or using large food containers reduces the amount of packaging waste generated.

(iii) **Toxicity reduction:** Source reduction reduces the amount of toxic constituents in products entering the waste stream and reduces the adverse environmental impacts of recycling or other waste management activities. For example, substitution of lead and cadmium in inks (solvent-based to water-based) and paints is a source reduction activity.

(iv) **Increased product lifetime:** Source reduction facilitates the use of products with longer lifetime over short-lived alternatives that are designed to be discarded at the end of their useful lives. Put differently, it encourages a product design that allows for repair and continued use rather than disposal. Manufacturing long-life tyres is a good example of increasing product lifetime.

(v) **Decreased consumption:** This refers to the reduced consumption of materials that are not reusable (e.g., using a reusable shopping bag instead of picking up plastic bags from the store). Consumer education about the materials that are difficult to dispose off or are harmful to the environment is essential. Buying practices can thus be altered (e.g., buying in bulk) to reflect environmental consciousness.

6.1.2 Implementation

There are several specific actions that can take place at the local level to encourage source reduction, some of which are given below:

(i) **Education and research:** Consumers, businesses, industries, schools, etc., can implement education and research activities to address the need for source reduction, its consequences, available choices, benefits and costs. Essentially, the aim of such education and research activities is to provide and develop information about source reduction goals, needs and methods and to elicit voluntary efforts from the public and private sectors to help bring about some specific changes. Some of the activities that reflect education and research to encourage source reduction include:

- forming stakeholder councils (industry, government agencies, etc.) to develop a source reduction message to the public and to carry out educational and research activities;
- exploring and developing funding sources, such as government grants, financial support for industries, private funding grant, direct tax and solid waste surcharge;
- developing media campaign for public outreach, including posters, conferences and forums concerning source reduction;
- developing curricula for schools and universities as well as organising a group of professionals with knowledge of source reduction and solid waste management.

(ii) **Financial incentives and disincentives:** Linking an economic benefit to the implementation of source reduction activities encourages source reduction. For example, financial disincentives represent additional costs to the waste producing activities that could be avoided through source reduction activities. The various measures that can be targeted at consumers and industries include the following:

- Tax credit or exemption may be given to companies and institutions that follow specific source reduction procedures for manufacturing and consuming.
- As variable waste disposal charges for garbage collection are generally taken for post-consumer solid waste, the fees or payment can be based on the number of garbage cans used, number of bags collected or the frequency of collection.
- Graduated fee structure for garbage collection can be based on quantity and frequency.
- Product disposal charges can be assessed either on the producers at the time of manufacture or on the consumers at the time of purchase. However, we need to note that though these charges can encourage source reduction on economic grounds and the funds generated from

the charges can be used to correct and reduce the impacts of product disposal, it is difficult to assess such charges effectively and efficiently.

(iii) **Regulation:** Although most regulation occurs at the national and state level, local authorities can participate in legislative activities in developing regulations that affect municipal SWM. It is possible, for example, to establish a programme to inform the consumers about environmental impacts, durability, reusability and recyclability of products as well as to declare source reduction as a top priority in SWM. Regulatory option of source reduction can include the following:

- Quantity control regulation, which encourages substitution of products that have the same function but pose less threat to human health and environment, through restrictions and bans.
- Product design regulation, which includes products that do not meet certain design criteria and could be subjected to quality control by sales tax and restrictions.

6.1.3 Monitoring

Monitoring facilitates the evaluation (i.e., efficacy and efficiency) of source reduction, the identification of possible source reduction measures and programme revisions and the obtaining of funds and resources for source reduction initiatives/ programmes. Monitoring should, therefore, be an integral part of a source reduction programme. We must, however, note that source reduction is more difficult to measure on a broad scale than other methods of SWM. When several waste reduction techniques are used simultaneously, for example, it is not easy to determine which portion of the diversion was due to source reduction.

Source reduction often results in substantial and measurable cost savings from activities such as waste collection, transportation and disposal, and direct savings. In addition, source reduction is cost-effective in decreasing pollution, purchase, use and regulatory compliance costs. It also reduces product and material use and disposal costs in the manufacturing process, making business operations more efficient. There is, however, some concern that source reduction might reduce economic growth by decreasing consumption. Nonetheless, it offers opportunities for economic gain by resource recovery and employment generation in source reduction programmes. Costs associated with source reduction must, therefore, form part of the monitoring process to ensure accountability.

Life Cycle Analysis

An environmental economic analysis will help us understand the trade-offs between source reduction, durability, recyclability, use of recycled material and other environmental benefits. Obviously, we choose the process that yields the greatest overall environmental benefits. Ideally, to assess and quantify these trade-offs, a life cycle analysis (LCA) – variously known as life cycle assessment, 'cradle to grave analysis' or 'econ-balance' – is required, which involves the evaluation of some aspects, particularly most often the environmental aspect, of a product through all stages of its life cycle.

LCA, in essence, takes a detailed look at all resources used and the products and by-products generated throughout the entire life of a product or process. The analysis begins with raw materials and energy acquisition, examines manufacturing and product fabrication, filling, packaging and distribution, and consumer use and reuse, and finally ends with an analysis of waste management.

Note that LCA represents a rapidly emerging family of tools and techniques designed to help in environmental management and long-term sustainable development (European Environment Agency, 1998) and the LCA procedures are being developed to assess the overall environmental impact of products and their packages. We will discuss LCA in detail later in Unit 10.

6.1.4 Evaluation

Before adopting source reduction policies, it is important that we develop a framework for evaluating various options. Some of the criteria to be considered in this regard are:

- Social and economic equity.
- Economic and administrative feasibility, efficiency and cost.
- Volume requirement and scarcity of materials and natural resources in product manufacture.
- Volume of product and its by-products that must be eventually disposed.
- Useful life, reusability and/or recyclability of a product.
- Priority of source reduction of more hazardous products to less hazardous ones.

Having studied the various aspects of source reduction, we will next deal with handling (i.e., reusing, recycling or disposing off) that portion of waste stream which cannot be reduced at the source. We will also explain how wastes can be recycled and utilised to derive from them optimum economic value.

LEARNING ACTIVITY 6.1

(i) Explain the need for source reduction in waste management.

(ii) List the criteria to be considered in the evaluation of various options before adopting a source reduction policy.

Note:

a) Write your answer in the space given below.

b) Check your answer with the one given at the end of this book.

6.2 SIGNIFICANCE OF RECYCLING

Recycling is perhaps the most widely recognised form of source reduction involving the process of separating, collecting, processing, marketing and ultimately using a material that would have otherwise been discarded. This form of source reduction, i.e., recycling, is similar to other forms, in that it:

- lessens reliance on landfills and incinerators;
- protects human health and the environment by removing harmful substances from the waste stream;
- conserves natural resources by reducing the demand for raw materials.

Recycling is one of the fundamental parts of the waste management plan. Although it alone cannot solve a community's municipal SWM problem, it can divert a significant portion of waste stream from disposal in landfill and combustion facilities. Recycling has a lot of direct and indirect significance for the society, and this can be grouped under the following three broad areas (UNCHS, 1994):

(i) **Economic significance:** Economic assessment of waste recycling is a difficult task as many of the beneficial environmental and social impacts of recycling are long-term and are intangible and, therefore, are difficult to quantify. Some of the short- and long-term economic benefits are:

- **Cost reduction:** Resource recovery through recycling of solid waste could be of interest to waste management authorities as a means of reducing the waste disposal cost. Any saving in waste management cost could be a significant incentive to the authorities to increase the coverage of service areas and improve the service level. They can save cost from fuel for transportation, operation and maintenance, and generate revenue by sale of recyclables, etc.
- **Employment:** Recycling of waste is a labour intensive activity, and its potential to ease the unemployment problem is high. Enhanced recycling activities, for example, can create an additional job market for skilled and unskilled workforce, and they can adapt to any of the occupations such as a labourer in recycling business or industry or a dealership.
- **Energy saving:** Use of recyclables in some industrial processes is known to consume less energy than the use of any other raw material. The reduction in energy consumption in one industry could mean its availability for some other industry in need.
- **Reduced health care costs:** Improved health and sanitary conditions in urban areas resulting from indirect benefits of waste recycling can reduce the investment in public health programme.
- **Saving costs for other public utilities:** Enhanced solid waste recycling practices can reduce the frequency of sewer clogging, blocking of natural watercourses and pollution of water bodies. This will benefit the concerned public utilities through reduced cost in cleaning sewers and improved public safety due to blocked sewers and narrowing of natural watercourses.

(ii) **Environmental and health significance:** The volume of waste is increasing rapidly because of population growth and economic development. The composition of waste is also changing, leading to waste production with more recyclables. At the same time, polluted waste fractions are increasing because of increasing complex processes being used in industries, and these contribute increasingly to environmental degradation. This notwithstanding, recycling helps, among others, in the following ways, to facilitate effective waste management:

- **Improved environment:** The environmental pollution may be due to inadequate SWM as well as due to its effect on other urban infrastructure. Recycling reduces the volume of waste that has to be finally dumped, and thereby causing reduction in pollution at the waste disposal sites. When there is reduction in volume of waste because of its increased reuse, different types of pollution (e.g., water, air and land) will get abated.
- **Natural resource conservation:** Industries with natural products as their raw material for production are depleting natural resources. Use of more and more recyclable solid wastes in industrial production will relieve the tremendous pressure on these precious resources. For example, recycling of waste paper means a lower demand for wood, which means less cutting of trees and an enhanced possibility for sustainable use of the forest. Using recyclable items in the production process would reduce the demand for energy as well.

(iii) **Social significance:** People engaged in waste collection activities are normally of low social and economic standing. This is especially true with scavengers, which is evident from persisting poor quality of their living and working conditions. Different groups of people engaged in waste recycling have a hierarchical social and economic status, in which, processors are at the top of the hierarchy followed by waste dealers and wholesalers, waste buyers and waste collectors in that very order, while scavengers are at the bottom. Although there is this social and economic hierarchy within the waste recycling business, the overall social esteem of waste recycling operators is low.

A formal recycling arrangement will help promote the social esteem of waste workers and facilitate their upward social mobility due to increased earning. In addition, the improved recycling activity will increase the economic value of the waste and will reduce waste scavenging activity providing opportunity for scavengers to switch to a more socially acceptable occupation. In short, institutionalised recycling programmes will help remove the stigma associated with waste scavenging and transform it to an economic enterprise.

Now that we have learnt the importance of recycling, which is more than just the separation and collection of post-consumer materials, let us plan a recycling programme, which indeed is the focus of Section 6.3.

✍ LEARNING ACTIVITY 6.2

Explain how recycling takes place in your locality and list its advantages?

Note:

a) Write your answer in the space given below.

b) Check your answer with the one given at the end of this book.

6.3 PLANNING OF A RECYCLING PROGRAMME

Numerous recycling options are available, and recycling programme development requires strategic planning. Planning for recycling involves understanding markets, assessing local expertise, setting goals and fostering public participation. An efficient recycling programme requires a systematic approach to all programme components, which are interrelated, and therefore, decisions about one must be made taking into consideration other components. As a successful recycling requires public participation, programmes must be designed keeping in view public convenience and support. The factors involved in the planning process include the following (EPA, 1989 and 1995):

(i) **Build local expertise:** Small projects help build local expertise in recycling and minimise the problems associated with poor planning. With small-scale projects, it is easy to compare and evaluate the programmes and techniques that are considered most successful within the community. When the time comes to develop a large-scale programme, there will be practical experience and an established decision-making framework, which will enhance the programme's success.

(ii) **Understand and develop a recycling market:** While planning for a recycling programme, it is important to find an outlet for the recyclable material. Market analysis is both a planning and ongoing activity, as even the most successful recycling programme can be severely affected by market fluctuations. Recycling programmes must, therefore, be designed with the flexibility to handle fluctuating markets and uncertain outlets for material.

(iii) **Foster public education and involvement:** Public participation is one of the most important factors deciding a programme's success. The public has a right and a responsibility to understand the full costs and liabilities of managing the waste they produce. A well-planned public education and involvement programme will foster public interest in recycling.

(iv) **Assess local waste stream:** Planning any recycling programme requires the knowledge of the local waste stream. Choosing the right material to recycle and designing the logistics of the programme are the important parts of the planning process.

(v) **Augment existing programme:** Recycling should augment the success that has been attained by other groups operating recycling programmes. This is very important for planning and success. Other programmes may be run by local volunteer organisations to raise funds or as a community service.

(vi) **Set goals and objectives:** Part of the planning process involves setting goals and objectives. The preliminary assessment of waste stream helps in deciding long-term goals for a community. Planning objectives may include determining the type of waste stream component that should be programmed, investigating the feasibility of the curbside (kerbside) programme, public outreach avenues, etc. The community will benefit from carefully developed achievable goals and objectives, and from an integrated approach to waste management (see Unit 10 for a detailed discussion of the integrated approach to waste management).

(vii) **Coordinate the programme:** Recycling programme is considered a public service. Therefore, local governments are required to ensure that all services are provided properly. Like any other public service, recycling programmes should be consistent, predictable, equitable and efficient.

(viii) **Evaluate the programme:** New programmes and technologies are evolving continuously, which make the planning for recycling an ongoing process.

This requires experiment and evaluation. Even the best recycling programmes experiment with new techniques to improve on their current efforts.

If recycling programmes are properly planned and implemented, they would then add to the overall municipal waste management activity.

LEARNING ACTIVITY 6.3

List the factors that you will consider while planning a recycling programme.

Note:

a) Write your answer in the space given below.
b) Check your answer with the one given at the end of this book.

6.4 RECYCLING PROGRAMME ELEMENTS

Recycling programmes are designed according to the needs and priorities of the communities. Elements of a recycling programme include source separation, curbside (kerbside) collection, material resource facilities and full stream processing. Recycling, generally, has a positive impact on other municipal waste management programmes. This may include a mix of strategies, ranging from simple, single material drop-off centres to large scale, centralised processing facilities. We will discuss some of the major elements (EPA, 1989 and 1995) in Subsections 6.4.1 to 6.4.8.

6.4.1 Source Separation

Source separation refers to the segregation of the recyclable and reusable materials at the point of generation. This requires that several designated materials be separated into their own specific containers, while other programmes use only two containers – one for the storage of mixed recyclables and the other for regular wastes. Source separation may be voluntary or mandated and is done in conjunction with several recycling programmes.

6.4.2 Drop-off/Buy-back

A *drop-off* programme requires residents to separate the recyclable materials and bring them to a specified drop-off or collection centre. However, we must take into account local conditions when designing a collection programme. For a community that does not provide a curbside pickup, for example, educating and encouraging citizens to deliver materials to a drop-off site may be all that is needed. A recycling centre can be established at the same location where residents deliver waste. Mobile recycling drop-off trailers can also be used. Drop-off recycling, however, is less convenient than curbside pickup. If a thorough educational and promotional effort is not made, drop-off programme tends to have lower participation rates than curbside collection.

Buy-back refers to a drop-off programme that provides monetary incentives to participate. In this type of programme, the residents are paid back for their recyclable material directly or indirectly through the reduction in collection and disposal fees. Establishing a buy-back centre (i.e., a place where recyclables are purchased) may help induce citizens to recycle. Some buy-back centres purchase some materials and accept others, depending on current market conditions. Private or public mobile buy-back operations can serve some areas, purchasing recyclables from small communities or from neighbourhoods of large metropolitan areas on a regular schedule.

6.4.3 Curbside Programme

In a curbside system, source separated recyclables are collected separately from regular refuse from the curbside, alley, or commercial facility. Curbside programmes vary greatly from community to community. Some programmes require residents to separate different materials that are stored in their own containers and collected separately. Other programmes use only one container to store recyclables or two containers, one for paper and the other for heavy recyclables (e.g., glass, aluminium, etc.).

6.4.4 Storage and Collection of Recyclables

Collection of source-separated materials is a necessary component of recycling programme. Establishing a collection system for source-separated materials will require more careful planning than regular trash collection. Some principles of sound recyclables storage and collection should be understood, while developing a programme, and these include:

- **Resident convenience:** The easier it is for residents to separate materials, the higher the participation and recovery rates will be.
- **Collection crew convenience:** The system should be convenient for collection crews. For example, loading and sorting activities should be as simple as possible.
- **Cost effectiveness:** Equipment and procedures must be designed to maximise collection crew and vehicle productivity.
- **Integrity of materials:** The storage and collection system should keep recyclables in the best shape possible. It should be properly handled, dry and contaminant-free.

How residents store recyclables in a household and curb has a direct impact on the success of a recycling programme. The storage container should provide a handy way to store materials until collected, and it must be easy for the collector to distinguish recyclables from garbage.

6.4.5 Collection Vehicles for Recycling

Collection vehicles that are designed specifically for collecting recyclables have several storage bins, which can be easily loaded and often equipped with automatic container-tipping devices. Although these modified vehicles may still be considered as options, a dedicated, closed-body collection vehicle for recycling with sufficient capacity offers such significant advantages as easy loading and unloading, flexible compartments and protection from weather. Of course, this warrants a substantial initial investment.

6.4.6 Processing Equipment for Recycling

Recycling involves a number of processing techniques and these processes require different equipments, many of which we discussed in Unit 5. However, some of the special equipments used in recycling are:

- **Balers:** Balers can be used to densify many types of materials including paper, cardboard, plastics and cans. Balers can improve space utilisation and reduce material transportation costs.
- **Can densifiers:** Can crushers are used to densify aluminium and steel cans prior to transport.
- **Glass crushers:** These are used to process glass fraction separated by colour and break it into small pieces. This crushed material is then called cullet, and can be reprocessed into new glass products.
- **Magnetic separators:** These are used to remove ferrous material from a mixture of materials.
- **Wood grinders:** These are chippers and are used to shred large pieces of wood into chips that can be used as mulch or as fuel.
- **Scales:** These are used to measure the quantity of materials recovered or sold.

6.4.7 Material Recovery Facilities (MRF)

MRF (pronounced 'murf') is a centralised facility that receives, separates, processes and markets recyclable material. It can be operated with both drop off and curbside programmes. The primary advantage of MRF is that it allows materials directly from the municipalities and processes them uniformly. It is generally designed to handle all type of recyclables. Implementation of MRF in a municipality depends upon a number of factors as follows:

- **Market demand:** When additional processing is required, MRF is more useful as buyers may have certain material specifications.
- **Separate collection:** In systems that require residents to separate their recyclables, intermediate separation and processing is required.
- **Number of different recyclables:** In general, a MRF will be more beneficial when a large number of different recyclables are collected.
- **Quantities of materials:** Because MRF involves substantial capital and operating costs (e.g., buildings, equipment and labour), it is expected to handle a significant amount of materials to justify its operation.

6.4.8 Full Stream Processing

This is a high technology separation technique, which processes all components of municipal waste. The materials recovered by this process tend to be of lower quality than those recovered or source separated in MRF because the former is a mix of various types of wastes. To achieve a better quality, the materials obtained through the full stream processing must be cleaned, which is a costly process. However, this technique remains attractive because it does not require source separation, and it is used in the following applications:

(i) **Refuse derived fuel (RDF) preparation:** In this application, it is used to extract the combustible portion of municipal waste.

(ii) **Municipal waste composting:** In this application, it is used to concentrate the compostable portion of municipal solid waste. Note that this is sometimes performed as part of RDF preparation.

(iii) **Material recovery:** In this application, it is used to recover and resell certain materials, and thereby making material recovery a recycling technology as well.

In full stream processing, depending on the facility design, the materials are separated either mechanically or by hand, and size and weight are the main characteristics used to separate the materials. For example:

- When the material is dumped, oversized materials such as furniture, etc., are removed.
- Rotating screens are used to separate materials of different sizes (small and large).
- Ferrous material is extracted using a magnet system.
- Air classifier is used to separate the lighter material.
- Light materials including plastic and paper are further processed into RDF.
- Heavy fraction is mechanically or manually sorted to recover saleable materials such as cardboard, etc.

LEARNING ACTIVITY 6.4

Explain the recycling programme elements adopted in your locality.

Note:

a) Write your answer in the space given below.

b) Check your answer with the one given at the end of this book.

6.5 COMMONLY RECYCLED MATERIALS AND PROCESSES

In this section, we will discuss some commonly recycled materials, the technologies used to recycle them and their specific environmental and economic impacts.

6.5.1 Paper and Cardboard

Paper and cardboard form the second biggest component of domestic waste after organic waste, and contribute to about 13% of the total domestic solid waste (UNCHS, 1994). Paper recycling is one of the most profitable activities and is practised extensively. It reduces the demand for wood and energy and helps solve littering problem in the city and around dumping site. It has an acceptable working condition and health risks are limited. Recovered paper and paper products are bought and sold through a well-established network of local processors and vendors who typically bale these materials for sale. Of late, paper mills have started buying directly from the collectors.

The paper industry is making a significant investment in manufacturing capacity for making paper and paper products with recycled content. Recovered paper is classified as newsprint, corrugated cardboard, mixed paper (including magazines, junk mail and cardboard), high-grade paper (white office paper, photocopying paper), and pulp substitute paper (usually mill scrap). Paper mills, the most common end users of recovered paper, use the material as a feedstock to manufacture recycled paper and paper products, such as newsprint, chipboard, craft linerboard, corrugating medium, roofing felt and tissue products. Shredded paper is used to make animal bedding, hydro mulch, moulded pulp products and cellulose insulation.

A typical input-output model of the paper processing technique consists of the following:

- **Material inputs:** Paper/cardboard scrap, magazines, newspapers, computer paper, wrapping paper, craft sacks, cartons, etc.
- **Product outputs:** Paper sheets, boxes, filter paper, mosquito mats (to absorb chemical repellent), merchandising packets, decorative items, etc.

And, the technology in the recycling of paper and cardboard involves the following processes:

(i) **Cardboard processing (semi-mechanical) plant:** In semi-mechanical plants, paper scrap is pulped in a beater machine. The paper pulp is spread on a rotating sieve and pressed mechanically. Cutting is done manually and after cutting and sun-drying, the cardboard is calendered and sheared into sheets, as illustrated in Fig. 6.1, from which boxes for shoes, sweets, etc., are made.

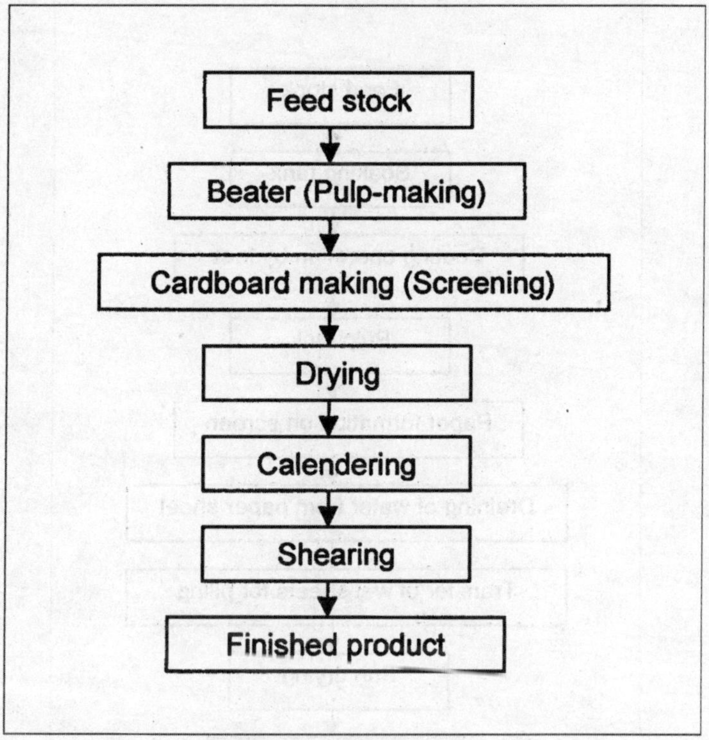

Figure 6.1 Cardboard Processing (Semi-Mechanical)

The various machineries used in the processes include:

- **Beater:** This consists of a masonry and concrete tank and a mild steel beater roll, driven by an electric motor.
- **Sieving screen:** This separates materials of different size.
- **Cardboard making unit:** This is a nylon conveyor screen driven by an electric motor.
- **Calendering machine:** This is used to finish the surface of paper and is electrically operated.

(ii) **Hand-made paper:** In hand-operated units, pulping and beating of paper is done manually in an open masonry or concrete tank. The material is sieved in a second tank and diluted with water to a specific consistency. A wooden framed screen is dipped in the tank in order to form an even layer of the wet pulp to a desired thickness (over the screen mat). The layer is skillfully removed and transferred from the mat and the sheets are sun dried on smooth walls or on other smooth surfaces. The dried sheets are then pressed and cut into required sizes. The calendering is performed in an electrically operated machine. Figure 6.2 illustrates the processing sequence in a hand-made paper unit.

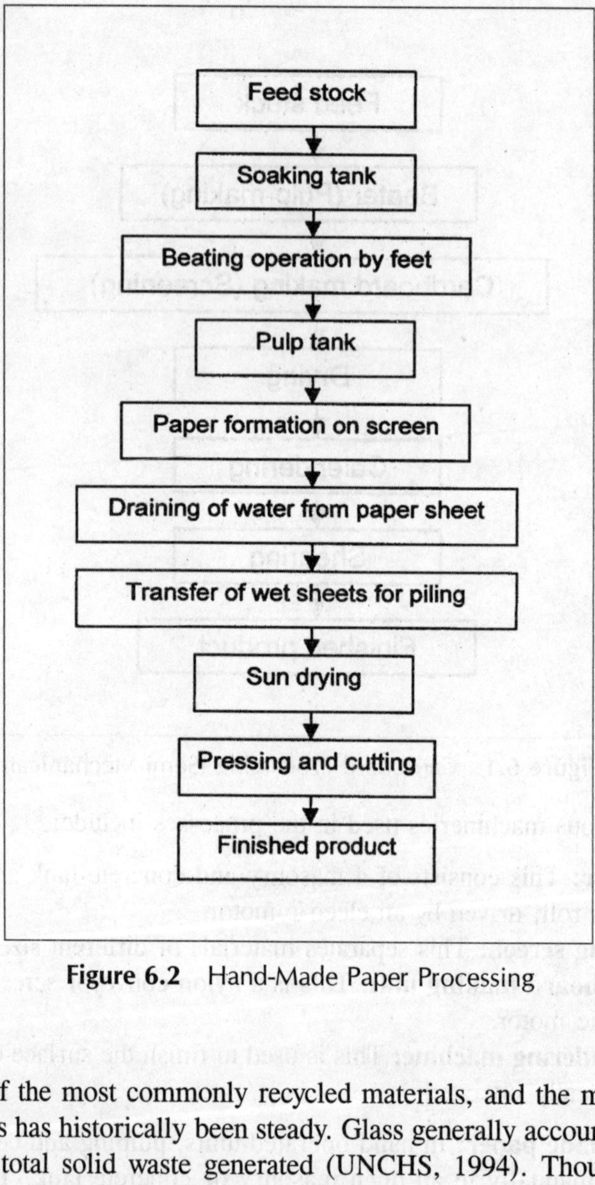

Figure 6.2 Hand-Made Paper Processing

6.5.2 Glass

Glass is one of the most commonly recycled materials, and the market for post-consumer glass has historically been steady. Glass generally accounts for 2.5% by weight of the total solid waste generated (UNCHS, 1994). Though it does not contribute to the environmental problem, glass does cause a serious problem of littering. The economic impacts are cost of waste collection and disposal, reduction in use of natural products and energy consumption.

Recycling of broken glass reduces the risk of diseases caused by cuts and wounds. Glass recycling is a labour intensive process and provides employment opportunity. Glass is typically broken for size reduction or crushed and ultimately sold to glass manufacturers as furnace-ready cullet after metal caps, rings, labels, etc., are removed.

Glass manufacturers purchase glass for reprocessing into new, clear, green and brown glass jars and bottles. The market for recovered glass has been strong and stable for brown and clear containers. Green glass, however, is seldom used to package goods domestically, so fewer companies produce glass of this colour. Alternative markets for glass include art glass, sandblasting, and industrial windowpane glass and fibreglass insulation. Although the glass industry has made a commitment to increase the demand for recovered glass, there is an important and pervasive market concern about the quality of material being produced by collection programmes and processing facilities. Recycling programme planners must address this concern for high-quality recovered glass as well as for other commodities.

Typically, while glass scrap is the material input, glass products are the product outputs in the glass recycling technique which involves a semi-mechanical process. This, we will study next.

Semi-mechanical process

The waste glass cullet is sorted according to colour and melted in an oven at 1400 °C. The oven used in a small production unit is locally made from fire clay and known as pot furnace. To improve the unbreakability of the glass product, chemicals such as soda ash, potassium carbonate, borax, lime, etc., are added to the cullet before melting. When the metal has completely melted, the temperature is raised to refine the glass. After refining, the temperature is lowered to carry out the moulding operation. After the glass takes the shape of the mould and hardens, unwanted portions are cut and removed, and sharp edges are smoothened in a machine. The product is then annealed by cooling and heating for three to four hours at 600–900 °C to reduce the brittleness of the glass. After quality inspection, defective glass products are returned to the recycling process. The approved products are packed for marketing. Figure 6.3 depicts the processes described above.

The following machinery is used in the process:

- **Furnace:** It is used to melt the glass cullet. The glass cullet is melted at 1400 °C (locally fabricated) and the product obtained is annealed at 600 °C (locally fabricated).
- **Semi-mechanical die:** It is used to cast the die. It uses both mechanically and manually operated moulds.
- **Air compressor:** It is used for blowing molten glass into dies.
- **Printing machine:** It is used for printing trademark.

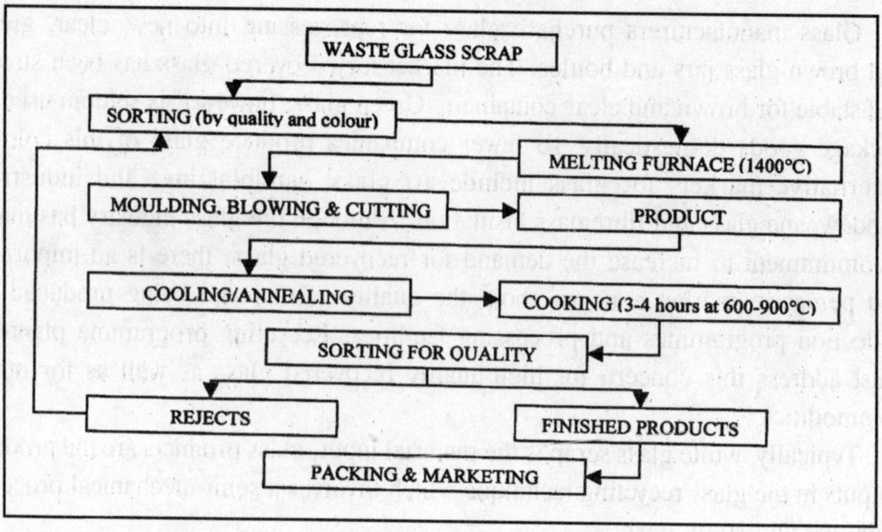

Figure 6.3 Glass Processing (Semi-Mechanical)

6.5.3 Metals

Ferrous metals like iron, steel, etc., and non-ferrous metals like aluminium, copper, zinc, lead, silver, etc., are some of the metals, which exist in the waste stream. On an average, metals account for 2% of total solid waste generated (UNCHS, 1994). Extraction of metals from natural ores depletes the mineral resources. Metals when dumped at landfill sites produce hazardous leachate with heavy metals in solution.

Using recycled metals substantially reduces operating costs of industries. Metal scrap is cheap and the energy consumption is lower when products are manufactured from scrap. The long-standing track record makes ferrous and non-ferrous metal market among the most stable of all recyclable materials. Ferrous and non-ferrous metals can be bought and sold through processors and vendors. Ferrous scrap includes household appliances, equipments, cans, and other iron and steel products. Non-ferrous scrap metals include aluminium, copper, lead, tin, etc. Both ferrous and non-ferrous metals can be prepared for sale through some combination of processing by flattening, baling, and shredding of the material. In some cases, processors melt the metal into ingots before selling it to end-use markets. Several foundries and steel mills have begun or expanded recycling efforts, and steel mini-mills also appear to be increasing their use of recovered steel in regions, which typically lack large mills.

The typical material inputs and product outputs in this industry are the following:

(i) **Material inputs:** Aluminium, brass, copper, zinc, tin, iron, steel, etc.
(ii) **Product outputs:** Sanitary and gas fittings, funnels, buckets and storage bins, reinforced steel bars, hand tools, etc.

Metal Processing

Now, we will discuss the process involved in recovering metals. Most of the recovered metals are processed by big industries. Ferrous metals are processed by iron industries to produce iron bars, channels, angles, etc. Local artisans process part of the ferrous metal in many cities. In a small-scale cottage industry, a particular kind of metal (ferrous or non-ferrous) is melted in a crucible in the coal furnace and the molten metal is cast into the desired mould to make ingots of required shapes and size. New and melted recycled metals are mixed together in a 3:1 ratio for better quality products. Ingots are sold to manufacturers to obtain different products from the metal. Local artisans heat the iron ingots in a coal furnace and beat them into different shapes for various kinds of tools and implements. Annealing hardens the cutting edges of the tools. The tools used in the process include coal burner, furnace (coal fired), moulding gadgets, dies (pattern) and auxiliary tools.

6.5.4 Plastic

These days, plastic is posing serious littering problem in cities and around collection points and dumping sites. With an average 8% by weight of the total amount of domestic waste, plastic is one of the major constituents in waste stream (UNCHS, 1994). Un-recycled plastic, when burned, contributes to green-house gases. The direct benefits of recycling plastic waste are reduction in the cost of raw material and energy saving. Plastic recycling also helps in employment generation along with reduction of volume transport and space requirements for dumping.

Most plastics are densified locally by flattening, baling, or granulating, and sold either to converters, where the resins are turned into pellets, or directly to domestic or export end users for remanufacture into products such as bottles, carpet and carpet backing, flower pots, and insulation material.

Post-consumer plastic-resin recycling technology has developed more rapidly than technologies for any other recovered material in the last half century. Only five to ten years ago post-consumer high-density polyethylene (HDPE) and polyethylene terephthalate (PET) plastics were vaguely considered recyclable. These two resins, especially HDPE milk jugs and bottles and clear PET plastics, now hold a stronger place in the market. End uses for recycled HDPE include non-food bottles, drums, toys, pipes, sheets and plastic pellet, and for PET include plastic fibres, injection moulding, non-food grade containers and chemicals. The recyclability of other resins, such as polystyrene, polyvinyl chloride, low-density polyethylene (LDPE), polypropylene and mixed plastic resins is making strides but much remains to be done. The input materials, the output products involved in the recycling of plastics are given below.

(i) **Material input:** Plastic scrap (thermoplastic and thermosetting).
(ii) **Product output:** Toys, boxes, slippers, shoes, pellets, buckets, cans, etc.

We will now discuss the processes involved in plastic recycling.

Plastic Processing

In plastic processing, the primary steps are sorting by colour and quality, and cutting and crushing the sorted material. The crushed product (granules) of plastic is melted, colour dyed and manually moulded into a cheaper product. The poly vinyl chloride plastic is blended with a specific colour dye in a mixing machine. The coloured material passes through an extruder machine to produce thick plastic strands. The strands are manually cut into lumps and these are used for manufacturing items either manually or mechanically, as shown in Fig. 6.4.

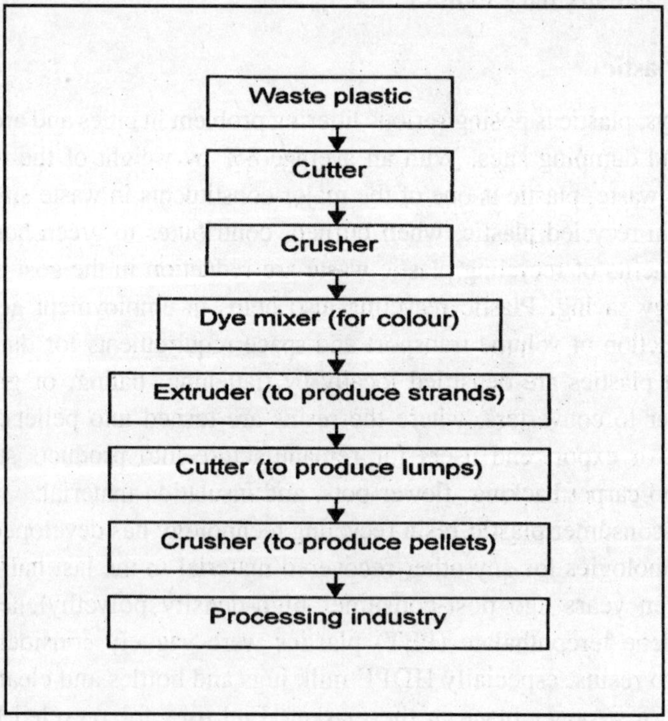

Figure 6.4 Plastic Processing (Raw Material for Plastic Product)

Note that in the manual process, the lumps are further chopped into smaller pieces and melted. The melted material is moulded into products such as shoe soles, toys and boxes. Machineries used in this process are electrically operated crusher, extruder, mixer and manually operated moulding machine (in which material is electrically heated at 359 °C,

6.5.5 Batteries and Tyres

Battery recycling is not only a response to market condition (i.e., price of lead) but also is important due to concern over the toxic compound including lead, cadmium and mercury present in many batteries. Like other materials, battery recycling depends largely on market conditions and requires consistent collection and processing. Household batteries come in a variety of types including alkaline, carbon, zinc, silver, nickel, cadmium, etc. Only those containing mercury and silver are marketed to end users, who extract metals. Automobiles use lead acid battery, which contains lead and sulphuric acid, both hazardous materials. Battery reprocessing includes breaking open the batteries, neutralising the acid, chipping the container for recycling and smelting the lead to produce recyclable lead.

Tyres represent a special challenge to solid waste and recycling programme managers. The use of chipped or shredded tyres as a source for fuel is growing. Electricity-generating facilities, pulp and paper mills and cement kilns are the most common processes using scrap tyres (EPA, 1989 and 1995).

✎ LEARNING ACTIVITY 6.5

State the end uses of recycled plastic.

Note:

a) Write your answer in the space given below.
b) Check your answer with the one given at the end of this book.

6.6 CASE STUDY: SOURCE REDUCTION AND RECYCLING IN BANGALORE

Source reduction, including reuse and recycling, can help reduce waste disposal and handling costs, because it avoids the costs of municipal composting, landfilling and combustion. Source reduction also conserves resources and reduces pollution, including greenhouse gases that contribute to global warming. Waste reduction, reuse and recycling, thus, play an important role in SWM. In what follows, we present the statistics on waste recovery and recycling done in Bangalore, India.

In Bangalore, 66% of the waste generated is collected for recovery, i.e., about 2373 tonnes per day. While 722 tonnes per day is reused, the rest (i.e., 1450 tonnes) goes for recycling. The agents involved in the collection and recovery of wastes in the city include waste pickers, IWB (i.e., itinerant waste buyer), middlemen (or intermediaries), the municipality and recycling units (both large and small). While the three agents in the informal sector and the municipality are directly involved in waste collection activities, the waste is processed by the recycling units, which receive recyclable waste from middlemen and municipality. Table 6.1 gives the statistics of waste recovered and collected by stakeholders in Bangalore.

Table 6.1 Waste Recovery and Collection by Stakeholders

Stakeholders	For reuse	For recycling		
		Waste pickers	IWB	Intermediaries
Households	101.7	74.5	60.4	57.7
Commercial establishments	350.8	227	0.0	24.3
• markets	112.7	200	0.0	8.8
• hotels	238.1	27	0.0	15.4
Institutes	0.0	9.6	0.0	61.3
• hospitals	0.0	1.0	0.0	0.0
• offices	0.0	8.4	0.0	0.0
• educational institutions	0.0	0.2	0.0	61.3
Industries	269.3	0.8	0.0	934.5
Total	721.8	312	60.4	1077.8

Of the 1450 tonnes collected for recycling, 1077.8 tonnes come from intermediaries, 60.4 come from IWB and 312 tonnes come from waste pickers. This amounts to 40% of the total waste (i.e., 3613 tonnes per day) generated.

SUMMARY

In this Unit, we discussed source reduction and recycling aspects, which precede the management of solid wastes. We began the Unit with an introduction to source reduction and its purpose, which is to reduce the amount of wastes generated at source and reduce their volume and/or toxicity before they become part of the solid waste stream. In this context, we explained how a source reduction programme could be successfully implemented and evaluated. We also said successful implementation requires education and research, financial incentives and disincentives and regulation and evaluation for various options using certain criteria.

Subsequently, we discussed recycling, which perhaps is the most widely recognised form of source reduction, and its significance. In addition, we explained the planning processes (i.e., understanding the market, identifying local expertise, setting goals and fostering public participation) and elements (i.e., source separation, curbside collection, material resource facilities and full stream processing) involved in a recycling programme. We then discussed commonly recycled materials such as paper, glass, metals, plastics and battery and tyres along with their processing techniques. Finally, we presented the source reduction and recycling programme carried out in Bangalore as a case.

SUGGESTED READINGS

Ali, M., Cotton, A. and Westlake, K. 1999. *Down to Earth*. Loughborough University, UK.

Areivala, S.J. 1971. *Solid Wastes Disposal in India*. Central Public Health Engineering Institute, Nagpur.

Arne Vesilind, William Worrel and Reinhart Debra. 2002. *Solid Waste Engineering*. Thomson Brooks/Cole, Singapore.

Attarwalla, F.A. 1993. *Solid Waste Management - A Manual*. All India Institute of Local Self-Government, Andheri (W), Bombay, India.

Flinotoff, R. 1984. *Management of Solid Wastes in Developing Countries*. WHO Regional Publications, Southeast Asia Series No 1.

Ogawa, H. 1989. *Selection of Appropriate Technology for Solid Waste Management in Asian Metropolises*, An International Journal of Regional Development Dialogue, UNCRD, Nagoya, Japan, Vol. 10, No.3, Autumn.

Ramachandra, T.V. and Saira Varghese, K. 2003. Exploring possibilities of achieving sustainability in solid waste management. *Indian Journal Environmental Health*, **45 (4):** 255-264.

Tchobanoglous, G., Theisen, H. and Eliassan, R. 1977. *Solid Wastes—Engineering Principles and Management Issues*. McGraw-Hill Book Company, New York.

Tchobanoglous, G., Theisen, H. and Samuel A. Vigil, 1993. *Integrated Solid Waste Management*, McGraw-Hill, Inc., New Delhi.

UNEP. 1996. *International Source Book on Environmentally Sound Technologies for Municipal Solid Waste Management (6)*, IETC, Osaka/Shiga.

http://ces.iisc.ernet.in/energy/SWMTR/TR85.html

REFERENCES

European Environment Agency. 1998. *Life Cycle Assessment (LCA): A Guide to Approaches, Experiences, and Information Sources*. Environmental Issues Series No 6.

Phelps, H.O., Heinke. G.W., Jonker, J.F., Ouano, E.A.R. and Vandecasteele, C. 1995. *Management of Solid Wastes*. UNESCO, Paris.

UNCHS. 1994. *A Reference Handbook for Trainers on Promotion of Solid Waste Recycling and Reuse in the Developing Countries of Asia*. UNCHS, Kenya.

US Environmental Protection Agency. 1989. *Decision-Maker's Guide to Solid Waste Management, Vol I*. Washington.

US Environmental Protection Agency. 1995. *Decision-Maker's Guide to Solid Waste Management, Vol II*. Washington.

UNIT 7

Recovery of Biological Conversion Products: Compost and Biogas

STRUCTURE

OVERVIEW

In Unit 6, we discussed source reduction to reduce the volume and/or toxicity of solid wastes. In that context, we explained recycling of waste and its significance, which is the most widely recognised form of source reduction. We also discussed some of the commonly recycled materials (such as paper, glass, metals, plastics and tyres) and their processing techniques. One of the environmentally sound and beneficial means of recycling organic materials is composting, and this we will discuss in this Unit.

Composting is the biochemical degradation of the organic fraction of solid waste material having a humus-like final product that could be used primarily for soil conditioning.

We will begin the Unit by discussing the physical, chemical and biological processes in composting along with their advantages. We will then touch upon the stages involved in composting, and subsequently, we will explain such composting technologies as windrow, aerated static pile, in-vessel and anaerobic composting. We will also discuss energy recovery through biogasification and deal with the factors affecting biogasification along with the types of digesters. We will close the Unit by discussing the environmental implications of composting and biogasification for land, water and air.

LEARNING OBJECTIVES

After completing this Unit, you should be able to:

- discuss the processes and stages involved in composting, and use appropriate composting technologies;
- explain biogasification, identify its advantages and select appropriate digesters for biogasification;
- assess the environmental effects of composting and biogasification;
- develop and operate a composting programme.

7.1 COMPOSTING

Biodegradation is a natural, ongoing biological process that is a common occurrence in both human-made and natural environments. In the broadest sense, any organic material that can be biologically decomposed is *compostable*. In fact, human beings have used this naturally occurring process for centuries to stabilise and recycle agricultural and human wastes. Today, however, composting is a diverse practice that includes a variety of approaches, depending on the type of organic materials being composted and the desired properties of the final product.

To derive the maximum benefit from the natural, but typically slow decomposition process (e.g., grass clippings left in the lawn or food scraps rotting in dustbins, etc.), it is necessary to control the environmental conditions during the composting process. The overall composting process can be explained as follows:

$$\textit{Organic matter} + O_2 + \textit{aerobic bacteria} \rightarrow CO_2 + NH_3 + H_2O + \textit{other end products} + \textit{energy}$$

Compost is the end product of the composting process. The by-products of this process are carbon dioxide and water. Compost is peaty humus, dark in colour and has a crumbly texture, an earthy odour, and resembles rich topsoil. Composts will not have any resemblance in the physical form to the original waste from which it was derived. High-quality compost is devoid of weed seeds and organisms that may be pathogenic to humans, animals, or plants. Cured compost is also relatively stable and resistant to further decomposition by microorganisms.

A complete composting process is shown in Fig. 7.1.

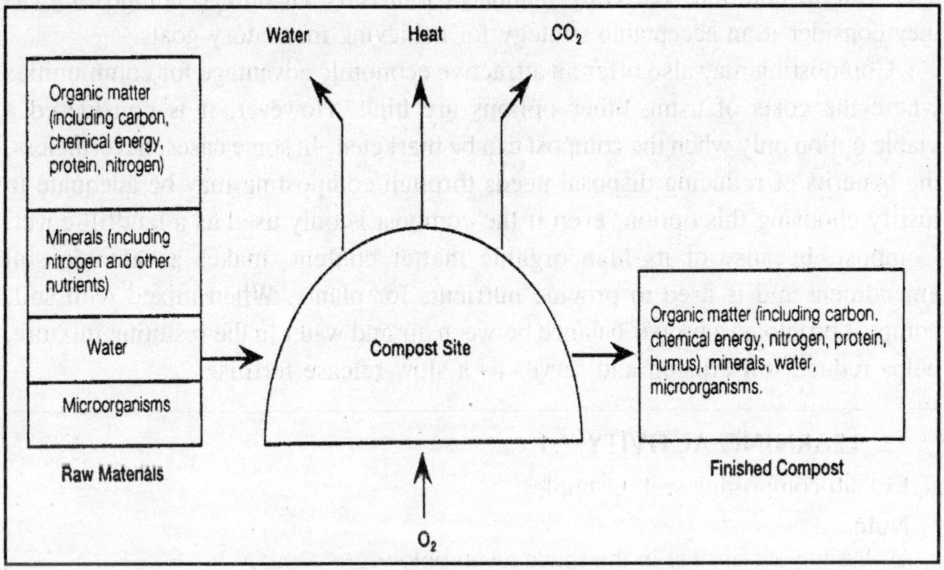

Figure 7.1 The Composting Process

As mentioned earlier, the composting process is an environmentally sound and beneficial means of recycling organic materials and not a means of waste disposal. And, it is important to view compostable materials as usable and not as waste requiring disposal. Before we discuss the composting process in detail, let us talk about the benefits of composting.

7.1.1 Benefits

Composting is one of the important components of solid waste management (SWM). It is a form of source reduction or waste prevention, as the materials are completely diverted from the disposal facilities and require no management or transportation. Community-yard trimming composting programme, source-separated organic composting and mixed municipal solid waste (MSW) composting constitute the various recycling processes. A major portion of municipal solid wastes in India contain up to 70% by weight of organic materials. In addition, certain industrial by-products – those from food processing, agricultural and paper industries – are mostly composed of organic materials.

Composting, being an organic material, can significantly reduce waste stream volume. Diverting such materials from the waste stream frees up landfill space needed for materials that cannot be composted. Composting owes its current popularity to several factors, including increased landfill tipping fees, shortage of landfill capacity and increasingly restrictive measures imposed by regulatory agencies. In addition, recycling mandates indirectly encourage composting, as they consider it an acceptable strategy for achieving mandatory goals.

Composting may also offer an attractive economic advantage for communities where the costs of using other options are high. However, it is considered a viable option only when the compost can be marketed. In some cases, nevertheless, the benefits of reducing disposal needs through composting may be adequate to justify choosing this option, even if the compost is only used as a landfill cover. Compost, because of its high organic matter content, makes a valuable soil amendment and is used to provide nutrients for plants. When mixed with soil, compost promotes a proper balance between air and water in the resulting mixture, helps reduce soil erosion and serves as a slow-release fertiliser.

✍ LEARNING ACTIVITY 7.1

Explain composting with examples.

Note:

a) Write your answer in the space given below.

b) Check your answer with the one given at the end of this book.

7.1.2 Processes

Several factors contribute to the success of composting (see Fig. 7.1), including physical, chemical and biological processes (EPA, 1989 and 1995). Understanding these processes, therefore, is necessary for making informed decisions, when developing and operating a composting programme.

Biological Processes

Microorganisms such as bacteria, fungi and actinomycetes as well as larger organisms such as insects and earthworms play an active role in decomposing the organic materials. As microorganisms begin to decompose the organic material, they break down organic matter and produce carbon dioxide, water, heat and humus (the relatively stable organic end product). This humus end product is compost.

Microorganisms consume some of the carbon to form new microbial cells, as they increase their population. They have, however, preferences on the type of organic material they consume. When the organic molecules they require are not available, they may become dormant or die. The chain of succession of different types of microbes continues, until there is little decomposable organic material left. At this point, the organic material that is remaining is termed *compost*. It is largely made up of microbial cells, microbial skeletons and by-products of microbial decomposition, and undecomposed particles of organic and inorganic origin. Decomposition may proceed slowly at first because of smaller microbial populations, but as populations grow in the first few hours or days, they rapidly consume the organic materials present in the feedstock. The number and kind of microorganisms are generally not a limiting environmental factor in composting non-toxic agricultural materials, yard trimmings, or municipal solid wastes, all of which usually contain an adequate diversity of microorganisms.

Under optimal conditions, composting proceeds through the following four phases (*http://www.weblife.org/humanure/Chapture3_9.html*)

(i) **Mesophilic, or moderate-temperature phase:** Compost bacteria combine carbon with oxygen to produce carbon dioxide and energy. The microorganisms for reproduction and growth use some of the energy and the rest is generated as heat. When a pile of organic refuse begins to undergo the composting process, mesophilic bacteria proliferate, raising the temperature of the composting mass up to 44 °C This is the first stage of the composting process. These mesophilic bacteria can include *E. coli* and other bacteria from the human intestinal tract, but these soon become increasingly inhibited by the temperature, as the thermophilic bacteria take over in the transition range of 44–52 °C.

(ii) **Thermophilic, or high-temperature phase:** In the second stage of the process, the thermophilic microorganisms are very active and produce heat. This stage can continue up to about 70 °C, although such high temperatures are neither common nor desirable in compost. This heating stage takes place rather quickly and may last only a few days, weeks, or months. It tends to remain localised in the upper portion of a compost pile where the fresh material is being added, whereas in batch compost, the entire composting mass may be thermophilic all at once. After the thermophilic heating period, the manure will appear to have been digested, but the coarser organic material will not be digested. This is when the third stage of composting, i.e., the cooling phase, takes place.

(iii) **Cooling phase:** During this phase, the microorganisms that were replaced by the thermophiles migrate back into the compost and digest the more resistant organic materials. Fungi and macroorganisms such as earthworms and sow bugs that break the coarser elements down into humus also move back in.

(iv) **Maturation or curing phase:** The final stage of the composting process is called curing, ageing, or maturing stage, and is a long and important one. A long curing period (e.g., a year after the thermophilic stage) adds a safety net for pathogen destruction. Many pathogens have a limited period of viability in the soil, and the longer they are subjected to the microbiological competition of the compost pile the more likely they will die a swift death. Immature compost can be harmful to plants. Uncured compost can, for example, produce phytotoxins (i.e., substances toxic to plants), robbing the soil of oxygen and nitrogen and contain high levels of organic acids.

Different communities of microorganisms predominate during the various composting phases. Initial decomposition is carried out by mesophilic microorganisms, which rapidly break down the soluble, readily degradable compounds. The heat they produce causes the compost temperature to rise rapidly. As the temperature rises above 40 °C, the mesophilic microorganisms become less competitive and are replaced by thermophilic (heat loving) ones. At temperatures of 55 °C and above, many microorganisms that are pathogenic to humans or plants are destroyed. Temperatures above 65 °C kill many forms of microbes and limit the rate of decomposition. Compost managers use aeration and mixing to keep the temperature below this point. During the thermophilic phase, high temperatures accelerate the breakdown of proteins, fats, and complex carbohydrates like cellulose and hemicellulose, the major structural molecules in plants. As the supply of these high-energy compounds become exhausted, the compost temperature gradually decreases and mesophilic microorganisms once again take over the final phase of curing or maturation of the remaining organic matter.

The composting process, therefore, should cater to the needs of the microorganisms and promote conditions that will lead to rapid stabilisation of the organic materials. While several microorganisms are beneficial to the composting process and may be present in the final product, some microbes are potential pathogens to animals, plants or humans. These pathogenic organisms must be destroyed in the composting process before the compost is distributed in the market place. Most of this destruction takes place by controlling the temperature of composting operations.

Chemical Processes

Several factors determine the chemical environment for composting. These include the presence of an adequate carbon food/energy source, a balanced amount of nutrients, the correct amount of water, adequate oxygen, appropriate pH and the absence of toxic constituents that could inhibit microbial activity (EPA, 1989 and 1995). Let us now describe each of these factors below:

(i) **Carbon/energy source:** For their carbon/energy source, microorganisms in the composting process rely on carbon in the organic material, unlike higher plants that rely on carbon dioxide and sunlight. The carbon contained in natural or human-based organic materials may or may not be biodegradable. The relative ease with which a material is biodegraded depends on the genetic makeup of the microorganisms present and the organic molecules that the organism decomposes.

Since most municipal and agricultural organics and yard trimmings contain an adequate amount of biodegradable forms of carbon, it is not a limiting factor in the composting process. As more easily degradable forms of carbon are decomposed, a small portion of the carbon is converted into microbial cells, and a significant portion is converted to carbon dioxide and lost to the atmosphere. As the composting process progresses, the loss of carbon results in a decrease in weight and volume of the feedstock. The less-easily decomposed forms of carbon will form the matrix for the physical structure of the final product.

(ii) **Nutrients:** Among the plant nutrients (i.e., nitrogen, phosphorus and potassium), nitrogen is of greatest concern, because it is lacking in some plant materials. The carbon-nitrogen ratio, which is established on the basis of available carbon rather than total carbon, is considered critical in determining the rate of decomposition. Leaves, for example, are a good source for carbon, and fresh grass, manure and slaughterhouse waste are the sources for nitrogen. To aid the decomposition process, the bulk of the organic matter should be carbon with just enough nitrogen. In general, an initial ratio of 30:1 (C: N or Carbon: Nitrogen) is considered ideal. Higher ratios tend to retard the process of decomposition, while ratios below 25:1 may result in odour problems. As the composting process proceeds and carbon is lost to the atmosphere, this ratio decreases.

Finished compost should have ratios of 15 to 20:1. Adding 3 – 4 kg of nitrogen material for every 100 kg of carbon should be satisfactory for efficient and rapid composting. To lower the carbon to nitrogen ratios, nitrogen-rich materials such as yard trimmings, animal manures, or bio-solids are often added. Adding partially decomposed or composted materials (with a lower carbon:nitrogen ratio) as inoculums may also lower the ratio. As the temperature in the compost pile rises and carbon:nitrogen ratio falls below 25:1, the nitrogen in the fertiliser is lost as gas (ammonia) to the atmosphere. The composting process slows, if there is not enough nitrogen, and too much nitrogen may cause the generation of ammonia gas, which can create unpleasant odours.

(iii) **Moisture:** Water is an essential part of all forms of life, and the microorganisms living in a compost pile are no exception. Since most compostable materials have lower than ideal water content, i.e., 50 to 60% of total weight, the composting process may be slower than desired, if water is not added. However, it should not be high enough to create excessive free flow of water and movement caused by gravity. Excessive moisture and flowing water form leachate, which creates potential liquid management problems including water and air pollution (e.g., odour). For example, excess moisture impedes oxygen transfer to the microbial cells, can increase the possibility of developing anaerobic conditions and may lead to rotting and obnoxious odours. Microbial processes contribute moisture to the compost pile during decomposition. Although moisture is added, it is also being lost through evaporation. Since the amount of water evaporated usually exceeds the input of moisture from the decomposition processes, there is generally a net loss of moisture from the compost pile. In such cases, adding moisture may be necessary to keep the composting process performing at its peak.

Controlling the size of piles can minimise evaporation from compost piles, as piles with larger volumes have less evaporating surface/unit volume than smaller piles. The water added must be thoroughly mixed so that the organic fraction in the bulk of the material is uniformly wetted and composted under ideal conditions. Properly wetted compost has the consistency of a wet sponge. Systems that facilitate the uniform addition of water at any point in the composting process are preferable.

(iv) **Oxygen:** Composting is considered an aerobic process. Decomposition can occur under both aerobic (requiring oxygen) and anaerobic (lacking oxygen) conditions. The compost pile should have enough void space to allow free air movement so that oxygen from the atmosphere can enter the pile and the carbon dioxide and other gases emitted can be exhausted to the atmosphere. To maintain aerobic conditions, in which decomposition occurs at a fast rate, the compost pile is mechanically aerated or turned frequently to expose the microbes to the atmosphere and to create more air spaces by fluffing up the pile.

A 10 to 15% oxygen concentration is considered adequate, although a concentration as low as 5% may be sufficient for leaves. While higher concentrations of oxygen will not negatively affect the composting process, circulation of an excessive amount of air can cause problems. For example, excess air removes heat, which cools the pile and also promotes excess evaporation. In other words, excess air slows down the rate of composting. Excess aeration is also an added expense that increases production costs.

(v) **pH:** The pH factor affects the amount of nutrients available for the microorganisms, the solubility of heavy metals and the overall metabolic activity of the microorganisms. A pH between 6 and 8 is considered optimum, and it can be adjusted upward by the addition of lime, or downward with sulphur, although such additions are normally not neccssary. The composting process itself produces carbon dioxide, which, when combined with water, produces carbonic acid, which could lower the pH of the compost. As the composting process progresses, the final pH varies, depending on the specific type of feedstock used and operating conditions. Wide swings in pH are unusual, since organic materials are naturally well buffered with respect to pH changes. Note that down swings in pH during composting usually do not occur.

What the foregoing discussion informs us is that the composition of material to be composted largely determines the chemical environment. In addition, several modifications can be made during the composting process to create an ideal chemical environment for the rapid decomposition of organic materials.

Physical Processes

The physical environment in the compost process includes factors such as temperature, particle size, mixing and pile size (EPA, 1989 and 1995). Each of these is essential for the composting process to proceed in an efficient manner as described below.

(i) **Particle size:** As composting progresses, there is a natural process of size reduction and the particle size of the material being composted is critical. Because smaller particles usually have more surface area per unit weight, they facilitate more microbial activity on their surfaces, which leads to rapid decomposition. The optimum particle size has enough surface area for rapid microbial activity and also enough void space to allow air to circulate for microbial respiration. The feedstock composition can be manipulated to create the desired mix of particle size and void space. For instance, through particle size reduction, we can increase the desired combination of void space and surface area for garden trimmings or municipal solid wastes. To improve the aesthetic appeal of finished composts, we can carry out particle size reduction, after the composting process is completed.

(ii) **Temperature:** Composting can occur at a range of temperatures, and the optimum temperature range is between 32 and 60 °C. Temperatures above 65 °C are not ideal for composting as thermal destruction of cell proteins kill the organisms. Similarly, temperatures below the minimum required for a group of organisms affect the metabolic activity (i.e., regulatory machinery) of the cells. Temperatures can be lowered, if required, by either increasing the frequency of mechanical agitation or using airflow throttling, temperature feedback control or blowers controlled with timers. Mixing or mechanical aeration also provides air for the microbes.

When compost is at a temperature greater than 55 °C for at least three days, pathogen destruction occurs. It is important that all portions of the compost material are exposed to such temperatures to ensure pathogen destruction throughout the compost. At these temperatures, weed seeds are also destroyed. After the complete pathogen destruction, temperatures may be lowered and maintained at slightly lower levels (51 to 55 °C). Attaining and maintaining 55 °C for three days is not difficult for in-vessel composting systems (see Subsection 7.1.4). However, to achieve pathogen destruction with windrow composting systems (see Subsection 7.1.4), the 55 °C temperature level must be maintained for a minimum of 15 days, during which time the windrows must be turned at least five times. The longer duration and increased turning are necessary to achieve a uniform pathogen destruction in the entire pile. Care should be taken to avoid contact between materials that have achieved these minimum temperatures and materials that have not.

(iii) **Mixing:** Mixing of feedstock, water and inoculants is important and is done by turning or mixing the piles after composting has begun. Mixing and agitation distribute moisture and air evenly, and promote the breakdown of compost clumps. Excessive agitation of open vessels or piles, however, can cool them and retard microbial activity.

LEARNING ACTIVITY 7.2

Explain the significance of various processes in composting.

Note:

a) Write your answer in the space given below.

b) Check your answer with the one given at the end of this book.

7.1.3 Stages

There are five basic stages involved in all composting practices, namely preparation, digestion, curing, screening or finishing, and storage or disposal. However, you must note that differences (among various composting processes) may occur in the method of digestion or in the amount of preparation and the finishing required. In choosing the type of process to be used and the amount of sophistication required, a number of criteria must be considered (Pavoni et al., 1975). In what follows in this subsection, we will study some of these criteria.

Preparation

The preparation phase of composting involves several steps, and these depend upon the sophistication of the plant and the amount of resource recovery practised. A typical preparation process, however, may include such activities as the sorting of recyclable materials, the removal of non-combustibles, the shredding, pulping, grinding and the adding of water sludge. Most plants utilise receiving equipment, which provides a steady flow of solid waste throughout the operation. Consistency of flow is accomplished by the use of storage hoppers and regulated conveyor system. After the solid wastes leave the receiving area, the bulky items, which could damage the grinders, are removed by hand. The separation of other non-compostable recyclable materials like glass, metal, rag, plastic, rubber and paper may be accomplished before or after comminution (i.e., reduction to small pieces or particles by pounding or abrading) by either hand or mechanical means. Those non-compostables, which are not salvaged, must be ultimately disposed off in a sanitary landfill.

After initial separation, most composting processes require the solid waste material to be reduced in particle size to facilitate handling, digestion and mixing of the product to provide more homogenous compost. The three major methods of comminution, which have been utilised in composting processes, include hammer mill, rasper/shredder and wet pulper, and the various equipment required are scales, receiving bins, conveyors, grinders and screens (see Unit 5 for details on equipment).

Since the refuse characteristics vary from one load to the next, a final step in the preparation phase of composting may be to adjust the moisture and nitrogen content of the solid waste to be composted. The optimum moisture content ranges from 45 to 55% of wet weight, while the optimum carbon to nitrogen ratio should be below 30%. The moisture and nutrient adjustments can be accomplished most efficiently through the addition of raw wastewater sludge. This increases the volume of composted material by 6 to 10%, in addition to accelerating the composting operation and producing a better final product in terms of nutrient contents.

Digestion

Digestion techniques are the most unique feature of the various composting processes and may vary from the backyard composting process to the highly controlled mechanical digester. Composting systems fall into the following two categories:

(i) windrow composting in open windrows;

(ii) mechanical composting in enclosed digestion chambers.

(See Subsection 7.1.4 for a discussion of the composting technologies.)

Curing

Organic materials, remaining after the first (rapid) phase of composting, decompose slowly, despite ideal environmental conditions. The second phase, which is usually carried out in windrows, typically takes from a few weeks to six months, depending on the outdoor temperatures, intensity of management and market specifications for maturity. With some system configurations, a screening step may precede the curing operation. During curing, the compost becomes biologically stable, with microbial activity occurring at a slower rate than that during actual composting. Curing piles may be either force-aerated or passive aerated with occasional turning. As the pile cures, the microorganisms generate less heat and the pile begins to cool.

Note that the cooling of piles does not always mean that the curing is complete. Cooling is merely a sign of reduced microbial activity, which can result from lack of moisture, inadequate oxygen within the pile, nutrient imbalance or the completion of the composting process. Curing may take from a few days to several months to complete. The cured compost is then marketed.

Screening or finishing

Compost is screened or finished to meet the market specifications. Sometimes, this processing is done before the compost is cured. One or two screening steps and additional grindings are used to prepare the compost for markets. During the composting operation, the compostable fraction separated from the non-compostable fraction, through screens, undergoes a significant size reduction. The non-compostable fraction retained on the coarse screen is sent to the landfill, while the compostable materials retained on finer screens may be returned to the beginning of the composting process to allow further composting. The screened compost may contain inert particles such as glass or plastic that may have passed through the screen. The amount of such inert materials depends on feedstock processing before composting and the composting technology used.

To successfully remove the foreign matter and recover the maximum compost by screening, the moisture content should be below 50%. Drying should be allowed only after the compost has sufficiently cured. If screening takes place before curing is complete, moisture addition may be necessary to cure the compost. The screen size used is determined by market specifications of particle size.

Storage or disposal

In the final analysis, regardless of the efficiency of the composting process, the success or failure of the operation depends upon the method of disposal. Even when a good market for compost exists, provision must be still made for storage. Storage is necessary because the use of composting is seasonal, with greatest demand during spring and winter. Therefore, a composting plant must have a 6-month storage area. For a 300 tonne per day plant, this will require about six hectares of storage area. Many composting operations combine their curing period with the storage period. The price at which compost can be sold depends on the benefit to be obtained from its use and the customers who are willing to make use of such benefits. Compost may be sold in bulk, finished or unfinished, as well as fortified with chemical fertilisers.

LEARNING ACTIVITY 7.3

Discuss briefly the basic steps involved in the composting practice.

Note:

a) Write your answer in the space given below.

b) Check your answer with the one given at the end of this book.

During our discussion of the stages in the composting process, we frequently referred to various technologies used in the composting process. We will discuss four general categories of composting technologies in Subsection 7.1.4.

7.1.4 Technologies

The composting technologies – windrow, aerated static pile, in-vessel composting and anaerobic processing (EPA, 1989 and 1995) – vary in the method of air supply, temperature control, mixing/turning of the material, time required for composting, and capital and operating costs. Besides these general categories of composting technologies, there are also some supporting technologies, which include sorting, screening, and curing.

Let us now discuss the four general categories of composting technologies.

Windrow composting

The windrow system is the least expensive and most common approach. Windrows are defined as regularly turned elongated piles, shaped like a haystack in cross section and up to a hundred metres or more in length. The cross-sectional dimensions vary with feedstock and turning equipment, but most municipal solid waste (MSW) windrows are 1.5 to 3 metres high and 3 to 6 metres wide as shown in Fig. 7.2 below.

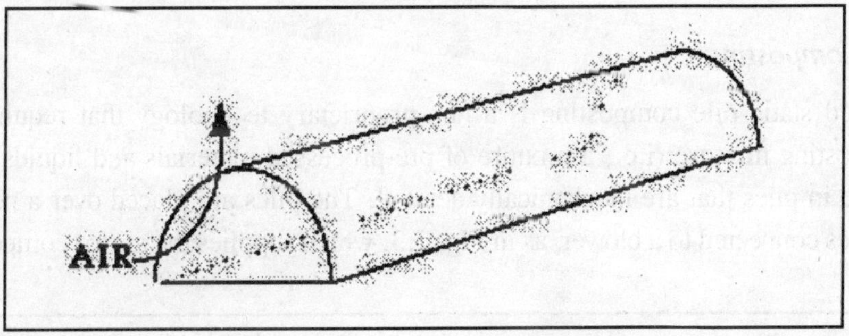

Figure 7.2 Windrow Composting

Windrows composed of MSW are usually required to be located on an impermeable surface, which greatly improves equipment handling under inclement weather conditions. The optimum size and shape of the windrow depends on particle size, moisture content, pore space and decomposition rate – all of which affect the movement of oxygen towards the centre of the pile.

Process control is normally through pile management, although forced aeration can also be used. Turning the pile re-introduces air into the pile and increases porosity so that efficient passive aeration from atmospheric air continues at all times. The windrow dimensions should allow conservation of the heat generated during the composting process and also allow air to diffuse to the deeper portions of the pile.

As mentioned earlier, windrows must be placed on a firm surface to turn the piles with ease. They may be turned as frequently as once per week, but more frequent turning may be necessary, if high proportions of bio-solids are present in the feedstock. Turning the piles also moves the materials from the pile surface to the core of the windrow, where they can undergo composting. Machines equipped with augers, paddles or tines are used for turning the piles. Some windrow turners can supplement piles with water, if necessary. When piles are turned, heat is released as steam to the atmosphere. If inner portions of the pile have low levels of oxygen, odours may result when this portion of the pile is exposed to the atmosphere. Piles with initial moisture content within the optimum range have a reduced potential for producing leachate. The addition of moisture from precipitation, however, increases this potential.

Any leachate or runoff created must be collected and treated or added to a batch of incoming feedstock to increase its moisture content. To avoid problems with leachate or runoff, piles can be placed under a roof, but doing so adds to the initial costs of the operation.

Aerated Static Pile Composting

Aerated static pile composting is a non-proprietary technology that requires the composting mixture (i.e., a mixture of pre-processed materials and liquids) to be placed in piles that are mechanically aerated. The piles are placed over a network of pipes connected to a blower, as in Fig. 7.3, which supplies the air for composting.

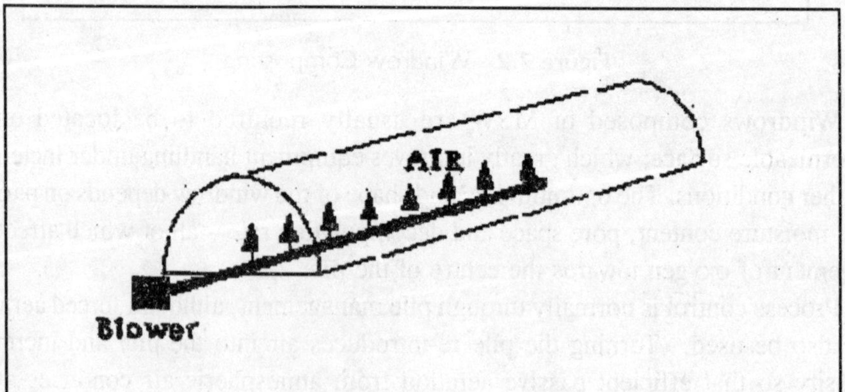

Figure 7.3 Aerated Static Pile Composting

Air can be supplied under positive or negative pressure. That is to say, the air supply blower either forces air into the pile or draws air out of it. Forcing the air into the pile generates a positive pressure system, while drawing air out of the pile creates a negative pressure. When the composting process is nearly complete, the piles are broken up for the first time since their construction. The compost is then taken through a series of post-processing steps. A timer or a temperature feedback system similar to a home thermostat controls the blowers.

Air circulation in the compost piles provides the needed oxygen for the composting microbes and prevents excessive heat build-up in the pile. Removing excess heat and water vapour cools the pile to maintain optimum temperature for microbial activity. A controlled air supply enables construction of large piles, which decreases the need for land. Odours from the exhaust air could be substantial, but traps or filters can be used to control them. The temperatures in the inner portion of a pile are usually adequate to destroy a significant number of the pathogens and weed seeds present. The surface of piles, however, may not reach the desired temperatures for destruction of pathogens because piles are not turned in the aerated static pile technology. This problem can be overcome by placing a layer of finished compost of 15 to 30 cms thick over the compost pile. The outer layer of finished compost acts as an insulating blanket and helps maintain the desired temperature for destruction of pathogens and weed seeds throughout the entire pile.

Aerated static pile composting systems have been used successfully for MSW, yard trimmings, bio-solids and industrial composting. Aerated static pile composting can also be done under a roof or in the open. Producing compost using this technology usually takes about 6 to 12 weeks. The land requirements for this method are lower than that of windrow composting.

In-vessel Composting System

In-vessel composting systems enclose the feedstock in a chamber or vessel that provides adequate mixing, aeration and moisture. Drums, digester bins and tunnels are some of the common in-vessel type systems. In-vessel systems vary in their requirements for pre-processing materials. For example, some require minimal pre-processing, while others require extensive MSW pre-processing. These vessels can be single- or multi-compartment units. In some cases, the vessel rotates, and in others, it is stationary and a mixing/agitating mechanism moves the material around. Most in-vessel systems are continuous-feed systems, although some operate in a batch mode. All in-vessel systems require further composting (curing) after the material has been discharged from the vessel.

A major advantage of in-vessel systems is that all environmental conditions can be carefully controlled to allow rapid composting. The material to be composted is frequently turned and mixed to homogenise the compost and promote rapid oxygen transfer. Retention times range from less than one week to as long as four weeks. These systems, if properly operated, produce minimal odours and little or no leachate. In-vessel systems enable exhaust gases from the vessel to be captured and are subjected to odour control and treatment. Some of the commonly used in-vessel systems are as follows (*http://www.adeq.state.ar.us/solwaste/*):

- **Vertical composting reactor:** It is generally over four metres high as illustrated in Fig. 7.4, and can be housed in *silos* or other large structures.

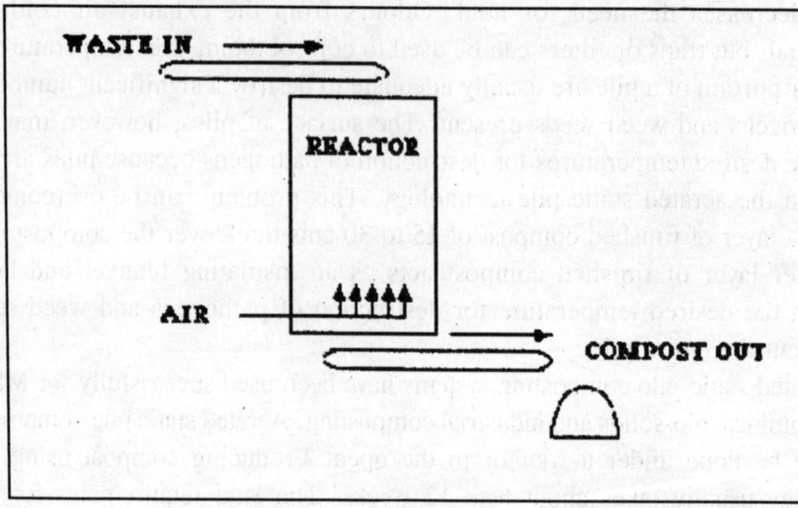

Figure 7.4 Vertical Composting Reactor

Organic material, typically fed into the reactor at the top through a distribution mechanism, moves by gravity to an unloading mechanism at the bottom. Process control is usually by pressure-induced aeration, where the airflow is opposite to the downward materials flow. The height of these reactors makes process control difficult due to the high rates of airflow required per unit of distribution surface area. Neither temperature nor oxygen can be maintained at optimal levels throughout the reactors, leading to zones of non-optimal activity. As with static pile composting, a stable porous structure is important in vertical reactors, which usually lack internal mixing. Tall vertical reactors have been successfully used in the sludge composting industry where uniform feedstock and porous amendments can minimise these difficulties in process control, but are rarely used for heterogeneous materials like MSW.

- **Horizontal composting reactors:** These avoid high temperature, oxygen and moisture gradients of vertical reactors by maintaining a short airflow pathway. They come in a wide range of configurations, including static and agitated, pressure and/or vacuum-induced aeration. Agitated systems generally use the turning process to move the material through the system in a continuous mode, while static systems require a loading and unloading mechanism. Material handling equipment may also shred to a certain degree, exposing new surfaces for decomposition, but excessive shredding may also reduce porosity. Aeration systems are usually set in the floor of the reactor and may use temperature and/or oxygen as control variables. Figure 7.5 illustrates a horizontal composting reactor.

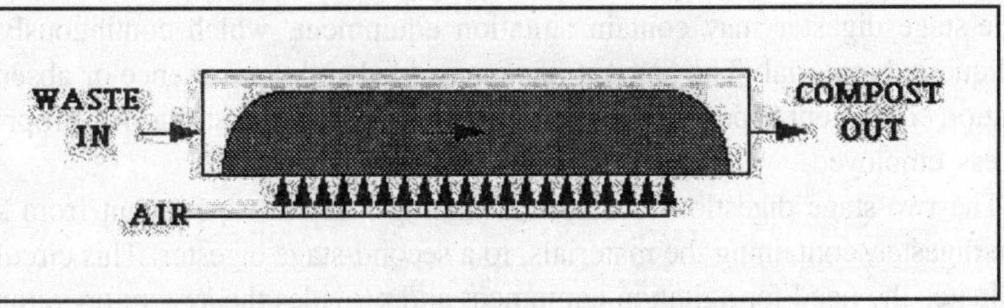

Figure 7.5 Horizontal Composting Reactor

Systems with agitation and bed depths less than two to three metres appear effective in dealing with the heterogeneity of MSW.

- **Rotating drum:** Rotating drum reactors (see Fig. 7.6) take the trade-off between reactor cost and compost residence time to an even further extreme than the horizontal or vertical in-vessel systems. These reactors, also known as digesters, retain the material for only a few hours or days.

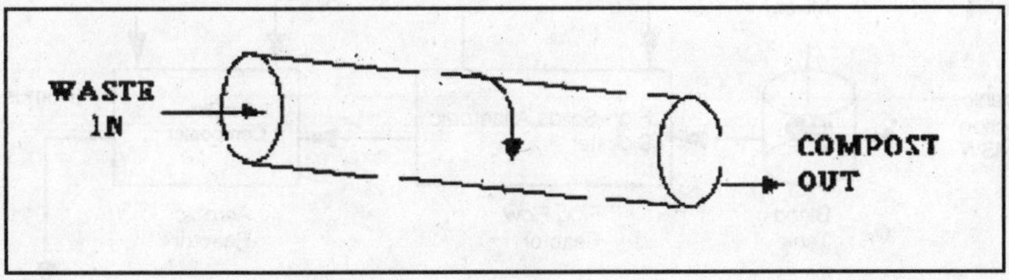

Figure 7.6 Rotating drum

While the tumbling action can help homogenise and shred materials, the short residence time usually means the processing is more physical than biological. While rotating drums can play an important role in MSW composting, they are normally followed by other biological processing, which may include in-vessel, static pile and/or windrow systems.

Anaerobic composting

In anaerobic processes, facultative bacteria break down the organic materials in the absence of oxygen and produce methane and carbon dioxide. Anaerobic systems, if configured efficiently, will generate sufficient energy in the form of methane to operate the process and have enough surpluses to either market as gas or convert to electricity. Conventional composting systems, on the other hand, need significant electrical or mechanical energy inputs to aerate or turn the piles. Several approaches are available for anaerobic digestion of feedstock.

Single-stage digesters contain the entire process in one airtight container. In this system, the feedstock is first shredded, and before being placed in the container, water and possibly nutrients are added to the previously shredded material. A single-stage digester may contain agitation equipment, which continuously stirs the liquefied material. The amount of water added and the presence or absence of agitation equipment depend on the particular research demonstration or proprietary process employed.

The two-stage digestion involves circulating a liquid supernatant from a first stage digester, containing the materials, to a second-stage digester. This circulation eliminates the need for agitation equipment and provides the system operator with more opportunity to carefully control the biological process. Figure 7.7 schematically illustrates an anaerobic digester with an aerobic compost curing.

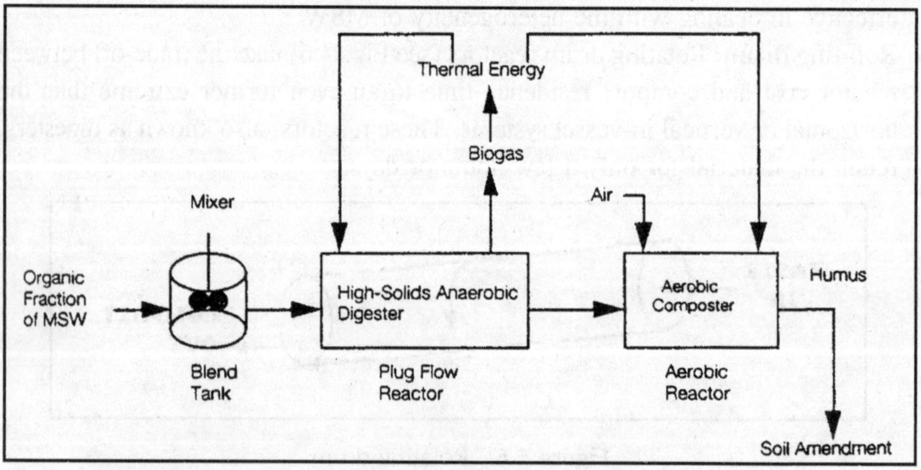

Figure 7.7 Anaerobic Digester with Aerobic Compost Curing

As digestion progresses, a mixture of methane and carbon dioxide is produced. These gases are continuously removed from both first- and second-stage digesters and are either combusted on-site or directed to off-site gas consumers. A portion of the recovered gas may be converted to thermal energy by combustion, which is then used to heat the digester. A stabilised residue remains, when the digestion process is completed. The residue is either removed from the digester with the mechanical equipment or pumped out as a liquid. It is chemically similar to compost, but contains more moisture. Conventional dewatering equipment can reduce the moisture content enough to handle the residue as a solid. The digested residue may require further curing by windrow or static pile composting. We will explain the process of anaerobic digestion and energy production through it in Section 7.2. But, first let us work out Learning Activity 7.4.

✍ LEARNING ACTIVITY 7.4

Explain anaerobic composting process.

Note:

a) Write your answer in the space given below.
b) Check your answer with the one given at the end of this book.

The Indian scenario

The first significant development in composting as a systemised process took place in India in 1925, when a process involving the anaerobic degradation of leaves, refuse, animal manure and sewerage were placed in pits. The materials were placed in layers and the pit wall conserved some of the heat of degradation, resulting in high temperature than when composting was carried out in the open. This process (often referred as Indore process) took approximately six months to produce usable compost. Following this, the Indian Council of Agricultural Research (ICAR) improved the method by laying down alternate layers of waste and sewage and this system (referred to as the Bangalore process) is still being used in India. In India, the high humid degradation that occurs in the land requires a large amount of humus for maintaining soil-fertility, and for that reason, composting is an ideal method for recycling organic wastes.

7.2 BIOGASIFICATION

Biogas is a mixture of gases composed of methane (CH_4) 40–70 vol.%, carbon dioxide (CO_2) 30–60 vol.%, other gases 1–5 vol.% including, hydrogen (H_2) 0–1 vol.% and hydrogen sulphide (H_2S) 0–3 vol.%. It originates from bacteria in the process of bio-degradation of organic material under anaerobic (without air) conditions. The natural generation of biogas is an important part of the biogeochemical carbon cycle.

Methanogens (methane producing bacteria) are the last link in a chain of microorganisms, which degrade organic material and return the decomposition products to the environment. In this process, biogas is generated, which is a source of renewable energy. As is the case with any pure gas, the characteristic properties of biogas are pressure and temperature dependency. It is also affected by moisture content. Well-functioning biogas systems can yield a whole range of benefits for their users, the society and the environment in general. Some of the important benefits are as follows:

- production of energy (heat, light, electricity);
- transformation of organic waste into high quality fertiliser;
- improvement of hygienic conditions through reduction of pathogens, worm eggs and flies;
- reduction of workload in firewood collection and cooking;
- environmental advantages through protection of soil, water, air and woody vegetation;
- micro-economical benefits through energy and fertiliser substitution, additional income generation and increasing yields of animal husbandry and agriculture;
- macro-economic benefits through decentralised energy generation, import substitution and environmental protection.

What we can deduce from the list above, is that biogas technology can substantially contribute to energy conservation and development, if the economic viability and social acceptance of biogas technology are favourable. In Subsections 7.2.1 to 7.2.3, we will discuss the various processes, environmental requirements and digesters associated with biogasification (*http://www3.gtz.de/gate/*).

7.2.1 Anaerobic Processing

Anaerobic processing of organic material is a two-stage process, where large organic polymers are fermented into short-chain volatile fatty acids. These acids are then converted into methane and carbon dioxide. The metabolic stages in biogasification are illustrated in Fig. 7.8.

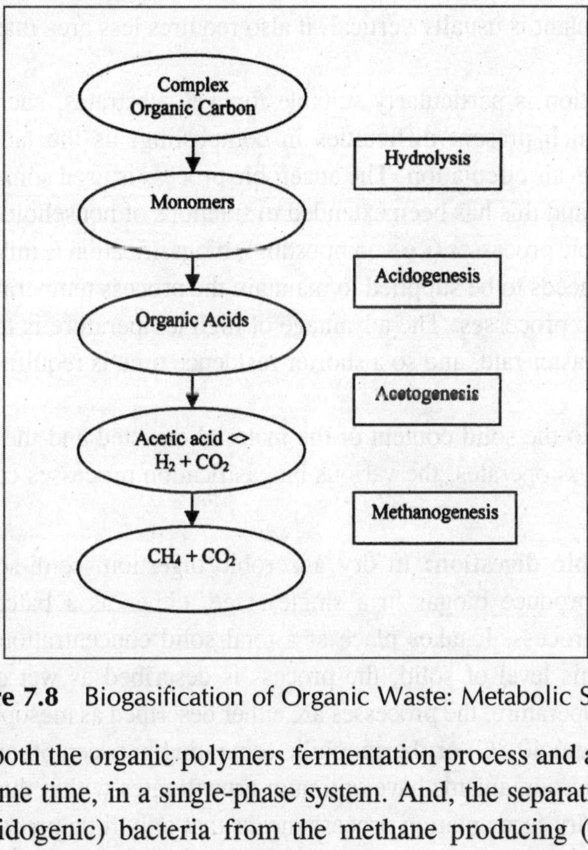

Figure 7.8 Biogasification of Organic Waste: Metabolic Stages

Note that both the organic polymers fermentation process and acid conversion occur at the same time, in a single-phase system. And, the separation of the acid-producing (acidogenic) bacteria from the methane producing (methanogenic) bacteria results in a two-phase system (McDougall et al., 2001).

The main feature of anaerobic treatment is the concurrent waste stabilisation and production of methane gas, which is an energy source. The retention time for solid material in an anaerobic process can range from a few days to several weeks, depending upon the chemical characteristics of solid material and the design of the biogasification system (e.g., single stage, two-stage, multi- stage, wet or dry, temperature and pH control).

In the absence of oxygen, anaerobic bacteria decompose organic matter as follows.

$$\textit{Organic matter} + \textit{anaerobic bacteria} \rightarrow CH_4 + CO_2 + H_2S + NH_3 + \textit{other end products} + \textit{energy}$$

The conditions for biogasification need to be anaerobic, for which a totally enclosed process vessel is required. Although this necessitates a higher level of technology than compared to composting, it allows a greater control over the process itself and the emission of noxious odours. Greater process control, especially of temperature, allows a reduction in treatment time, when compared to composting. Since a biogas plant is usually vertical, it also requires less area than a composting plant.

Biogasification is particularly suitable for wet substrates, such as sludges or food waste, which present difficulties in composting, as the lack of structural material restricts air circulation. The anaerobic process is used sometimes to digest sewage sludge, and this has been extended to fractions of household solid waste. In contrast to aerobic processes (i.e., composting), biogasification is mildly exothermic. Thus, the heat needs to be supplied to maintain the process temperature, especially for thermophilic processes. The advantage of high temperature is that the reaction will occur at a faster rate, and so a shorter residence time is required in the reactor vessel.

According to the solid content of the material digested and the temperature at which the process operates, the various biogasification processes can be classified as under:

- **Dry anaerobic digestion:** In dry anaerobic digestion, semi-solid wastes are digested to produce biogas in a single stage, either as a batch process or a continuous process. It takes place at a total solid concentration of over 25%, and below this level of solid, the process is described as wet digestion. With regard to temperature, the processes are either described as mesophilic (operating between 30 and 40 °C) or thermophilic (operating between 50 and 65 °C), and anaerobic microorganisms have optimum growth rates within these temperature ranges. The dry fermentation process means that only little process water has to be added or heated, which favours thermophilic operation. No mixing equipment is necessary, and crust formation is not possible due to the relatively solid nature of the digester contents. This anaerobic process usually takes between 12 and 18 days, followed by several days in post-digestion for residue stabilisation and maturation.

- **Wet anaerobic digestion:** In its simplest form, this process consists of a single stage in a completely mixed mesophilic digester, operating at a total solid content of around 3–8%. To produce this level of dilution, a considerable amount of water has to be added and heated, and then removed after the digestion process. This method is routinely used to digest sewage sludge, and animal and household wastes. The single-stage wet process can suffer from several practical problems such as the formation of a hard scum layer in the digester and difficulty in keeping the content completely mixed.

  The major problem with the single-stage process is that the different reactions in the process cannot be separately optimised. The acidogenic microorganisms grow fast and lower the pH of the reaction mixture, whereas the methanogens, which grow slowly, have a pH optimum around 7.0. The development of the two-stage digestion process solves this problem as hydrolysis and acidification occur in the first reactor vessel, kept at a pH of around 6.0 and methanogenesis occurs in the second vessel, operated at a pH of 7.5–8.2. The whole process can run with a retention time of 5 to 8 days. (We will discuss single-stage and two-stage digesters in Subsection 7.2.2.)

Maturing or refining

The residues of both wet and dry biogasification processes require extensive maturing under aerobic conditions. However, we can considerably reduce this period through effective aeration. The maturation processes facilitate the release of entrapped methane, elimination of phytotoxins (i.e., substances that are harmful for plant growth, such as volatile organic acids) and reduce the moisture content to an acceptable level. These residues contain a high level of water – even the dry process residue contains around 65% water.

Filtering or pressing can reduce excess water, and further drying can be achieved using waste heat from the gas engine, if the biogas is burnt onsite to produce electricity. The digested residue, initially anaerobic, will also contain many volatile organic acids and reduced organic materials. These need to be matured aerobically to oxidise and stabilise the compounds, in the process similar to the maturation of aerobic compost, prior to sale as compost or disposal as residue. Odour production is measured as the total amount of volatile organics produced per tonne of bio-waste during composting and the final aerobic maturation after anaerobic digestion.

Factors affecting biogasification

As with composting, a number of environmental factors (Phelps et al., 1995) influence biogasification, some of which are listed below.

- **Temperature:** A temperature range of about 25–40 °C (mesophilic) is generally optimal. It can be achieved without additional heating, thus being very economical. In some cases, additional energy input is provided to increase temperature to 50–60 °C (thermophilic range) for greater gas production. Often digesters are constructed below ground to conserve heat.
- **pH and alkalinity:** pH close to neutral, i.e., 7, is optimum. At lower pH values (below 5.5), some bacteria carrying out the process are inhibited. Excess loading and presence of toxic materials will lower pH levels to below 6.5 and can cause difficulties. When pH levels are too low, stopping the loading of the digester and/or use of time is recommended. The presence of alkalinity between 2500 and 5000 mg/L will provide good buffering against excessive pH changes.
- **Nutrient concentration:** An ideal C:N ratio of 25:1 is to be maintained in any digester. It is an important parameter, as anaerobic bacteria need nitrogen compound to grow and multiply. Too much nitrogen, however, can inhibit methanogenic activity. If the C:N ratio is high, then gas production can be enhanced by adding nitrogen, and if the C:N ratio is low, it can be increased by adding carbon, i.e., adding chicken manure, etc., which reduces the possibility of toxicity. Anaerobic digestion not only breaks down plant materials into biogas, but also releases plant nutrients, such as nitrogen (N), potassium (K) and phosphorous (P), and converts them into a form that can be easily absorbed by plants.
- **Loading**: When any digester is designed, the main variable to be defined is the internal volume. The digester volume is related to, as Fulford (1998) shows, two other parameters, and these are feed rate (Q, measured in m^3/day) and hydraulic loading or retention time (R, measured in days). The feed rate (Q) is given by the mass of total solid (m, kg) fed daily, divided by the proportion of total solid (TS) in the mixed slurry (as summing the density of feed is 1000 kg/m^3).

  $$Q = \frac{m}{TS \times 1000} \text{ or } Q = \frac{m}{TS\% \times 10} \text{ m}^3\text{/day}$$

  The retention time (R) of any digester is given by the volume of the digester pit (V, m^3), divided by the volume of the daily feed (Q, m^3/day).

  $$R = \frac{V}{Q} \text{ days}$$

  The loading rate, r (kg. VS/m^3/day) of a digester is defined as the mass of volatile solids added each day per unit volume of digester. This is related to mass feed rate:

$$r = \frac{m \times VS}{Q} \text{ or } r = \frac{m \times VS\%}{Q \times 100} \text{ kg VS/m}^3\text{/day}$$

The typical values for the loading rate are between 0.2 and 2.0 kg VS/m^3/day.

- **Effect of toxins:** The main cause of biogas plants receiving flak is the presence of toxic substance. Chlorinated hydrocarbons, such as chloroforms and other organic solvents, are particularly toxic to biogas digestion. If the digester has been badly poisoned, it may be difficult to remove the toxins without removing most of the bacteria. In that case, the digester must be emptied, cleaned with plenty of water and refilled with fresh slurry.

7.2.2 Types of Digesters

In the anaerobic digestion process, the organic matter in mixtures of sludges is converted biologically, under anaerobic conditions, to a variety of products including methane (CH_4) and carbon dioxide (CO_2). The process is carried out in an airtight reactor, where wastes in the form of sludges are introduced continuously or intermittently and retained in the reactor for varying periods. The microbiology of anaerobic digestion and the optimum environmental considerations for the microorganisms can be achieved by selecting the proper type of digester.

Against this background, we explain below the operation and physical facilities for anaerobic digestion in single-stage digester (standard rate and high rate) and two-stage digester (Tchobanoglous and Burton, 1996), which generally operate in a mesophilic range, i.e., between 30 and 38 °C.

(i) **Standard rate single-stage digester:** In a single-stage digester, the untreated waste sludge is directly added to the zone, where the sludge is actively digested and the gas is being released. The sludge is heated by means of an external heat exchanger.

As the gas rises to the surface, it lifts sludge particles and other minerals such as grease oil and fats, ultimately giving rise to the formation of scum layer. As a result of digestion, the sludges stratify by forming a supernatant layer above the digesting sludge and become more mineralised, i.e., the percentage of fixed solid increases. Due to stratification and lack of mixing, the standard rate process is used principally for small installations. Detention time for standard rate processes vary from 30 to 60 days.

Figure 7.9 illustrates a single-stage digester.

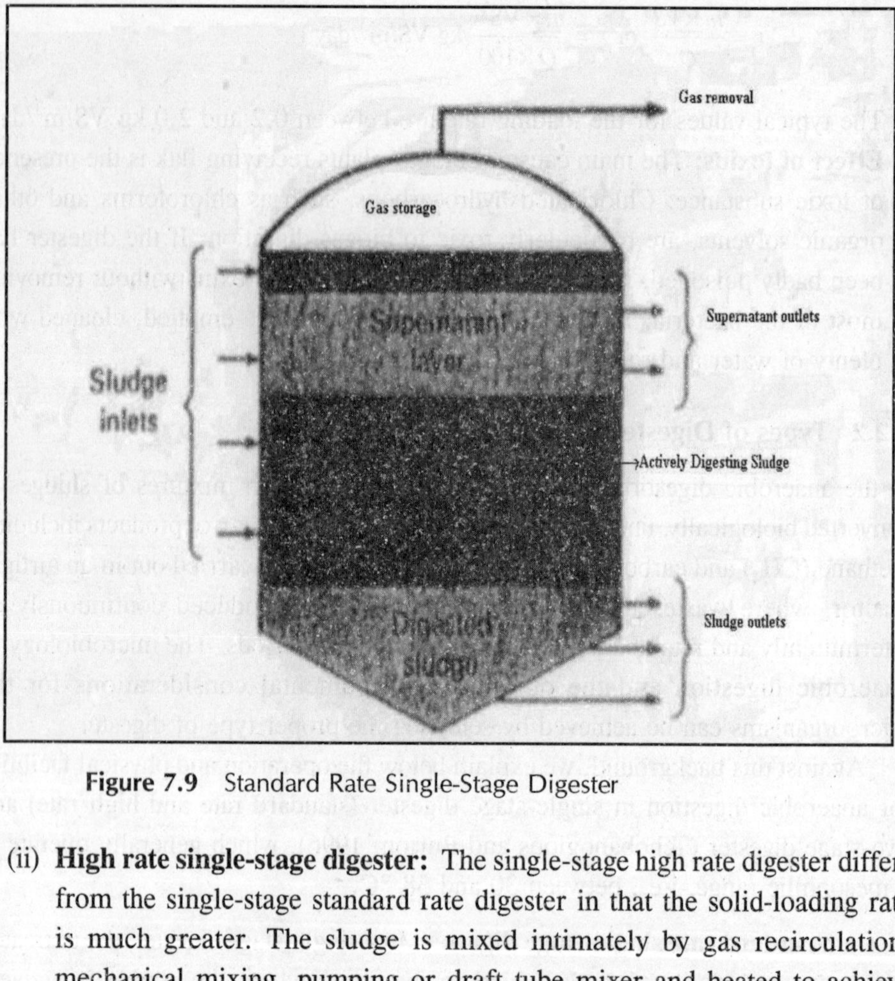

Figure 7.9 Standard Rate Single-Stage Digester

(ii) **High rate single-stage digester:** The single-stage high rate digester differs from the single-stage standard rate digester in that the solid-loading rate is much greater. The sludge is mixed intimately by gas recirculation, mechanical mixing, pumping or draft tube mixer and heated to achieve optimum digestion rates.

With the exception of higher loading rates and improved mixing, there are only a few differences between standard rate and high rate digester. The mixing equipment should have a greater capacity and should reach the bottom of the tank, which should be deeper to aid the mixing process. Digestion tank may have fixed roof or floating covers along with gasholder facility, which provide extra gas storage capacity. The required detention time for a high rate digestion is typically 15 days or less.

Figure 7.10 illustrates a high rate single-stage digester.

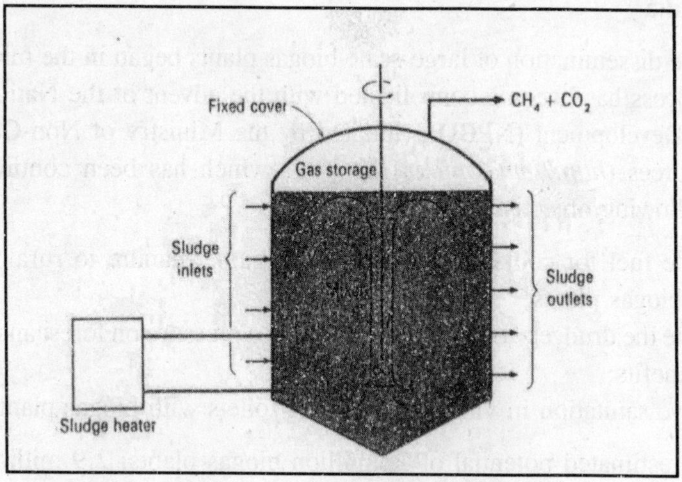

Figure 7.10 High Rate Single-Stage Digester

(iii) **Two-stage digester:** The combination of the two digesters, mentioned above, is known as a two-stage digester. The first stage digester is a high rate complete mix digester used for digestion, mixing and heating of waste sludge, while the primary function of a second stage is to separate the digested solid from the supernatant liquor, and in the process, additional digestion and gas production may occur. The tanks are made identical, in which case either one may be the primary digester. They may have fixed roofs or floating covers along with gasholder facility. Figure 7.11 below illustrates a two-stage digester.

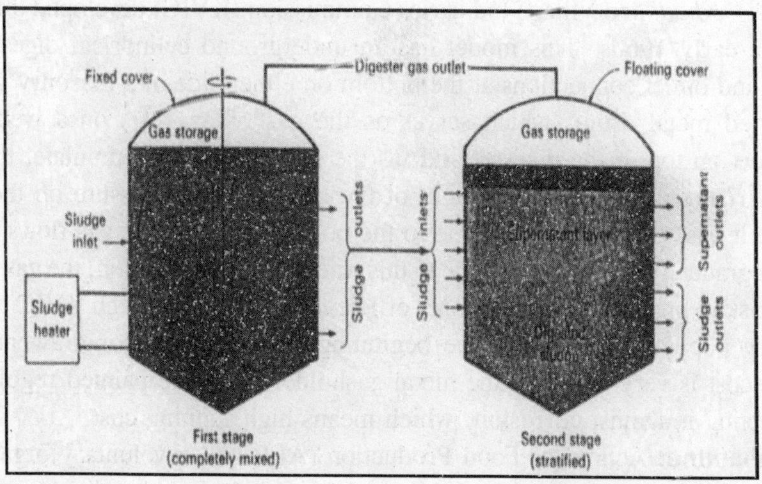

Figure 7.11 Two-Stage Digester

7.2.3 Biogas Plants in India

In India, the dissemination of large-scale biogas plants began in the mid-seventies, and the process has become consolidated with the advent of the National Project on Biogas Development (NPBD), initiated by the Ministry of Non-Conventional Energy Sources (*http://mnes.nic.in*) in 1981, which has been continuing since, with the following objectives:

- to provide fuel for cooking purposes and organic manure to rural households through biogas plants;
- to mitigate the drudgery of rural women, reduce pressure on forest and accentuate social benefits;
- to improve sanitation in villages by linking toilets with biogas plants.

Against the estimated potential of 12 million biogas plants, 2.9 million families, and 2700 communities, institutional and night soil-based biogas plants were set up by December 1999 (*http://www.teriin.org/renew/tech/biogas/*). This is estimated to have helped in saving three million tonnes of fuel wood per year and manure containing nitrogen equivalent to 0.7 million tonnes of urea. There are various biogas plant models approved by NPDB for implementation. All these models are based on the basic designs available: floating metal drum type, fixed masonry dome type, and FLEXI, a portable model made of rubberised nylon fabric, which has been approved for promotion in hilly and other terrains. Some of the NPDB approved models are as follows:

- **KVIC floating drum:** Based on *Gramalakshmi III*, the first workable prototype of a biogas plant - created by Jashbhai Patel, a Gandhian worker in the early 1950s - *Khadi* and Village Industries Commission (KVIC) developed this model in the early 1960s. This model has an underground cylindrical digester with inlet and outlet connections at the bottom on either side of a masonry wall. An inverted metal drum, which serves as the gasholder, rests on a wedge type support on top of the digester and, as the gas begins to accumulate, the drum starts rising in height. The weight of the drum applies pressure on the gas to make it pass through the pipeline to the point of use. As the gas flows out, the drum gradually moves down. Due to this smooth two-way motion, the gas remains at constant pressure, which ensures efficient use of gas. Though KVIC has been a very popular model since the beginning, it has two major drawbacks: the plant cost is very high and the metal gasholder has to be painted regularly for protecting it against corrosion, which means high running cost.
- **Deenbandhu:** Action for Food Production (AFPRO), a voluntary organisation based in New Delhi, developed this model in 1984. *Deenbandhu* has probably

been the most significant development in the entire biogas programme of India, as the cost of the plant was nearly 50% less than that of the KVIC plant, and this brought the biogas technology within the reach of even the poorer sections of the population. The cost reduction was achieved by minimising the surface area through joining the segments of two spheres of different diameters at their bases. This structure acts as a digester, and pressure is exerted on the slurry, which is pushed into a displacement chamber. Once the gas is drawn from the outlet, the slurry again enters the digester. The brick masonry dome, which is fixed, requires skilled workmanship and quality material to ensure no leakage.

- **Pragati:** The United Socio-Economic Development and Research Programme (UNDARP), a Pune-based NGO (non-government organisation) developed this model, which is a combination of KVIC and *Deenbandhu* designs. In this model, the lower part of the digester is semi-spherical in shape with a conical bottom. However, instead of a fixed dome, it has a floating drum acting as gas storage chamber. The spread of *Pragati* model has been confined mainly to the state of Maharashtra.
- **KVIC plant with ferro-cement digester:** The Structural Engineering Research Centre, Ghaziabad, has developed the ferro-cement digester design in order to overcome the problems encountered in construction of traditional models of biogas plants. In this digester, layers of thin steel wire mesh, distributed throughout, are impregnated with rich mortar. Ferro-cement (i.e., concrete made of welded mesh, sand and cement) as a building material offers several advantages like 10 to 15% reduction in cost over KVIC digesters, usage of locally available material, less labour and little or no maintenance. The only limitation though is ferro-cement, which is vulnerable to damage due to the impact of pointed loads, can be easily repaired.
- **KVIC plant with fiber reinforced plastic (FRP) gasholder:** FRP is used in place of metal in the floating drum gasholder. Contact Moulding Process, a technique of moulding without the application of external pressure, is adopted to manufacture the FRP. It employs one of the less expensive types of moulds, resulting in lower cost of the plant. The major advantage of FRP is its resistance to corrosion, which saves the recurring expenditure on painting the drum.
- **FLEXI:** Developed by Swastik Rubber Products Ltd., Pune, this is a portable model in which the digester is made of rubberised nylon fabric. The model is particularly suitable for hilly areas, where high transportation cost of construction materials, such as cement and bricks substantially increases the cost of installing the regular type of biogas plants.

Components of biogas plant

A typical biogas plant has the following components:

- a digester in which the slurry (e.g., dung mixed with water) is fermented;
- an inlet tank used to mix the feed and let it into the digester;
- a gas holder/dome in which the generated gas is collected;
- an outlet tank to remove the spent slurry;
- distribution pipeline(s) to carry the gas into the kitchen; and
- a manure pit, where the spent slurry is stored.

✍ LEARNING ACTIVITY 7.5

Do you suggest biogasification approach for MSW? Give reasons.

Note:

a) Write your answer in the space given below.

b) Check your answer with the one given at the end of this book.

So far, we studied the importance and benefits of composting and biogasification. In Section 7.3, we will examine the effects they may have on the environment.

7.3 COMPOSTING AND BIOGASIFICATION: ENVIRONMENTAL EFFECTS

The compost produced and slurry from biogas plants is often used as a soil amendment in a variety of applications, after ascertaining the quality of the product. Since the product is used for a variety of purposes, contaminants in the compost could be detrimental to the environment (EPA, 1989 and 1995). This is particularly true when one considers that the environmental pollutants are easily minimised through simple treatment procedures (i.e., waste segregation, proper turning and sufficient composting time). In the main, composting facilities of solid waste with manure, sewage sludge and residue from processing create some significant environmental considerations including the following:

(i) **Water impacts:** Water runoff from garden waste composting facilities could contain large concentration of nutrients (i.e., nitrates and phosphorous), volatile organics and metals that cause algal blooms in the nearby surface water. Retention basins may be used at facilities to limit water runoff. Water impacts are not generally expected to be serious at yard waste composting facilities. Municipal waste composting, sludge composting and co-composting involve a large amount of potential contaminants, and water impact could be greater at these facilities. Leachate from MSW compost facilities can contain high concentration of nutrients (e.g., nitrate and phosphorous), volatile organics and metals. Leachate could affect both surface and ground water.

(ii) **Land impacts:** At garden waste composting facilities, soil may become more acidic, due to the presence of certain leaves and pine needles in the compost pile. Nitrogen depletion may also occur, which can be limited by proper turning. MSW and co-composting facilities carry the potentially harmful impacts of acids, and organic and metal contamination. Again careful pre-processing to divert much of the potentially hazardous materials from the compost facility is an important quality control procedure.

(iii) **Odour pollution:** Odour is one of the most frequent problems at composting facilities. Frequent turning of compost piles has proven to be effective in limiting odour problems. When in-vessel systems are used, odour control devices (e.g., air scrubber, etc.) can minimise the problem.

(iv) **Health impact:** The primary public health concerns associated with composting operation result from:

- drinking contaminated water;
- toxins in the finished product (applied on land);
- pathogens.

Nitrate contamination of drinking water can affect the oxygen-carrying capacity of the blood in infants, e.g., blue baby syndrome (methemoglobinemia) and in elderly, but again under proper composting conditions, this risk can be minimised. Insects and vermin can spread pathogens. Note that because of the high temperature achieved during the normal composting operations, pathogens found in manure, sewage sludge or municipal waste are usually destroyed. There are risks to workers as well (e.g., aggravation of respiratory problem). However, proper training and health monitoring as well as proper apparel and equipment can minimise these risks.

✍ LEARNING ACTIVITY 7.6

List the environmental implications of composting.

Note:

a) Write your answer in the space given below.

b) Check your answer with the one given at the end of this book.

SUMMARY

In this Unit, we discussed the benefits and environmental effects of composting, a biochemical degradation of the organic fraction of solid waste material having humus-like final product that could be used primarily for soil conditioning, and biogas. We began the Unit by describing some of the benefits of composting. We then explained the physical, chemical and biological processes in composting. In that context, we said that while the mesophilic, thermophilic, cooling and curing phases are involved in the biological processes, the carbon/energy content, nutrients, oxygen, moisture and pH influence the chemical process and particle size, temperature and mixing affect the physical process. We also discussed the stages (preparation, digestion, curing or finishing and storage and disposal) and the technologies (windrow, aerated static pile, in-vessel and anaerobic) of composting processes. Following this discussion, we described biogasification, another process for recovery of biological conversion, and anaerobic processing, including the factors affecting biogasification, types of digesters and the biogas plants in India. Finally, we closed the Unit by examining the impact of composting and biogasification on the environment.

SUGGESTED READINGS

Arne Vesilind, William Worrel and Reinhart Debra. 2002. *Solid Waste Engineering*. Thomson Brooks/Cole, Singapore.

Areivala, S.J. 1971. *Solid Wastes Disposal in India*. Central Public Health Engineering Institute. Nagpur.

Tchobanoglous, G., Theisen, H. and Samuel A. Vigil. 1993. *Integrated Solid Waste Management*, McGraw-Hill, Inc., New Delhi.

Ramachandra, T.V. and Saira Varghese, K. 2003. Exploring possibilities of achieving sustainability in solid waste management. *Indian J. Environmental Health*, **45 (4)**: 255-264.

Environmental Management in Developing Countries. 1995. *Waste Management Vol II*, Institute for Scientific Co-operation, Tubingen.

Flinotoff, R. 1984. *Management of Solid Wastes in Developing Countries*. WHO Regional Publications, Southeast Asia Series No 1.

http://ces.iisc.ernet.in/energy/SWMTR/TR85.html

REFERENCES

Fulford, D. 1998. *Running A Biogas Programme: A Handbook*. Intermediate Technology Publications, UK.

McDougall, F.R., White, P.R., Franke, M. and Hindle, P. 2001. *Integrated Solid Waste Management: A Life Cycle Inventory*. Blackwell Science, UK.

Pavoni, J.L., Heer, J.E. and Hegerty, D.J. 1975. *Handbook of Solid Waste Disposal*. Van Nostrand Reinhold Company, New York.

Phelps, H.O., Heinke. G.W., Jonker, J.F., Ouano, E.A.R. and Vandecasteele, C. 1995. *Management of Solid Wastes*. UNESCO, Paris.

Tchobanoglous, G. and Burton, L.F. 1996. *Wastewater Engineering*. Tata McGraw-Hill Book Company, New Delhi.

US Environmental Protection Agency. 1989. *Decision-Maker's Guide to Solid Waste Management, Vol I*. Washington.

US Environmental Protection Agency. 1995. *Decision-Maker's Guide to Solid Waste Management, Vol II*. Washington.

http://www.adeq.state.ar.us/solwaste/compost.htm#Intro

http://www.teriin.org/renew/tech/biogas/models.htm#drum

http://www.weblife.org/humanure/chapter3_9.html

UNIT 8

Incineration and Energy Recovery

STRUCTURE

OVERVIEW

In Unit 7, we discussed the recovery of energy through composting and biogasification. Yet another means of energy recovery is thermal treatment of solid wastes, which is also regarded as pre-treatment of waste prior to final disposal. Thermal treatment includes both the burning of mixed MSW (municipal solid waste) in incinerators and the burning of selected parts of the waste stream as fuel. In this Unit (i.e., Unit 8), we will deal with incineration and energy recovery. We will begin the Unit with an introduction to the process and objectives of incineration. Then, we will take up the issues that are to be considered while planning an incineration facility. We will, subsequently, discuss various incineration technologies such as mass burning system, refuse derived fuel (RDF) system, modular incineration and fluidised bed incineration. Then, we will explain energy generation, i.e., generation of steam and electricity, and cogeneration of steam and electricity. Emission of air pollutants being a major concern of incineration facilities, we will also discuss the various gaseous pollutants and their control measures (equipment). Finally, we will discuss the environmental impacts of incineration on land, water and aesthetics.

LEARNING OBJECTIVES

After completing this Unit, you should be able to:

- discuss incineration processes;
- list the objectives of incineration;
- plan an incineration facility;
- explain various incineration technologies;
- identify emissions from incinerators and their control;
- estimate the energy generation potential of wastes;
- assess the environmental impacts of incineration.

8.1 INCINERATION: AN INTRODUCTION

Incineration is a chemical reaction in which carbon, hydrogen and other elements in the waste mix with oxygen in the combustion zone and generates heat. The air requirements for combustion of solid wastes are considerable. For example, approximately 5000 kg of air is required for each tonne of solid wastes burned. Usually, excess air is supplied to the incinerator to ensure complete mixing and combustion and to regulate operating temperature and control emissions. Excess air requirements, however, differ with moisture content of waste, heating values and the type of combustion technology employed. The principal gas products of combustion are carbon dioxide, carbon monoxide, water, oxygen and oxides of nitrogen.

Many incinerators are designed to operate in the combustion zone of 900–1100 °C. This temperature is selected to ensure good combustion, complete elimination of odours and protection of the walls in the incinerator. Incinerator systems are designed to maximise waste burn out and heat output, while minimising emissions by balancing the oxygen (air) and the three "Ts", i.e., time, temperature and turbulence. Complete incineration of solid wastes produces virtually an inert residue, which constitutes about 10% of the initial weight and perhaps a larger reduction in volume. The residue is generally landfilled.

The incineration facility along with combustion of waste emits air pollutants (i.e., fine particulate and toxic gases), which are an environmental concern and, therefore, their control is necessary. Other concerns relating to incineration include the disposal of the liquid wastes from floor drainage, quench water, scrubber effluents and the problem of ash disposal in landfills because of heavy metal residues. By optimising the combustion process, we can control the emission of combustible, carbon-containing pollutants (EPA, 1989 and 1995). Oxides of nitrogen and sulphur, and other gaseous pollutants are not considered a problem because of their relatively smaller concentration.

Having introduced you to incineration, we will touch upon the combustion of various elements of waste materials and the objectives of incineration in Subsections 8.1.1 and 8.1.2, respectively.

8.1.1 Combustion of Waste Material

Table 8.1 shows the major elements that constitute solid wastes and the end products of combustion.

Table 8.1 Major Elements of Solid Wastes

Elements		**End Products (Gases)**
Carbon (C)		Carbon dioxide (CO_2)
Hydrogen (H)		Water (H_2O)
Oxygen (O)	Combustion Process	
Nitrogen (N)		Nitrogen (N_2)
Sulphur (S)		Sulphur dioxide (SO_2), other gaseous compounds and ash

Table 8.2 gives the information about several components of solid waste mixtures on the basis of proportion.

Table 8.2 Ultimate Analysis of Combustible Component

	Percent by Weight (dry basis)					
Component	**Carbon**	**Hydrogen**	**Oxygen**	**Nitrogen**	**Sulphur**	**Ash**
Food waste	48.0	6.4	37.6	2.6	0.4	5.0
Paper	43.5	6.0	44.0	0.3	0.2	6.0
Cardboard	44.0	5.9	44.6	0.3	0.2	6.0
Plastic	60.0	7.2	22.8	—	—	10.0
Textile	55.0	6.6	31.2	4.6	0.15	2.5
Rubber	78.0	10.0	—	2.0	—	10.0
Leather	60.0	8.0	11.6	10.0	0.4	10.0
Garden trimmings	47.8	6.0	38.0	3.4	0.3	4.5
Wood	49.5	6.0	42.7	0.2	0.1	1.0
Dirt, ash, brick, etc.	26.3	3.0	2.0	0.5	0.2	68.0

Source: Tchobanoglous et al. (1977, 1993)

In case energy values in KJ/kg or BTU/lb are not available, we can calculate them approximately from the data in Table 8.2 above and the Dulong formula given below.

$$\textit{Energy value}\ (\text{BTU/lb}) = 145.4\ \text{C} + 620\ (\text{H} - 1/8\ \text{O}) + 41\ \text{S} \tag{8.1}$$

where C, H, O and S are in percent by weight (dry basis) and can be converted to KJ/kg by ***BTU/lb* × *2.326* = *KJ/kg***

8.1.2 Incineration Objectives

The purpose of incineration is to combust solid wastes to reduce their volume to about one-tenth, without producing offensive gases and ashes (Phelps et al., 1995). That is to say, incineration of solid wastes aims at the following (McDougall et al., 2001):

- **Volume reduction:** Depending on its composition, incineration reduces the volume of solid wastes to be disposed off by an average of 90%. The weight of the solid wastes to be dealt with is reduced by 70–75%. This has both environmental and economic advantages since there is less demand for final disposal to landfill, as well as reduced costs and environmental burdens due to transport, if a distant landfill is used.
- **Stabilisation of waste:** Incinerator output (i.e., ash) is considerably more inert than incinerator input (i.e., solid wastes), mainly due to the oxidation of the organic components of the waste stream. This leads to a reduction of landfill management problems (since the organic fraction is responsible for landfill gas production) and the organic compounds present in landfill leachate.
- **Recovery of energy from waste (EFW):** This represents a valorisation method, rather than just a pre-treatment of waste prior to disposal. Energy recovered from burning the wastes is used to generate steam for use in on-site electricity generation or export to local factories or district heating schemes. Combined heat and power plants increase the efficiency of energy recovery by producing electricity as well as utilising the residual heat. Solid waste incineration can replace the use of fossil fuels for energy generation. As a large part of the energy content of solid wastes comes from truly renewable resources (e.g., biomass), there should be a lower overall net carbon dioxide production than that from burning fossil fuels, since carbon dioxide is absorbed in the initial growing phase of the biomass.
- **Sterilisation of waste:** This is of primary importance in the incineration of clinical or biomedical waste. Incineration of solid wastes will also ensure destruction of pathogens prior to final disposal in a landfill.

The prevalence of incineration practice and the actual approach that it takes vary across regions and reflect the relative importance attached to the different objectives discussed. Volume reduction for both environmental and economic reasons and sterilisation of waste has historically been the important objectives of incineration. Due to the growing concern over the production of landfill gas and the organic compounds in leachate from landfill receiving untreated waste, it is likely that the future will see more emphasis on using incineration for stabilising wastes for subsequent landfilling. Landfill gas and leachate arise principally from the organic fraction of solid wastes, which will be effectively converted to gases and mineralised ash by incineration.

✍ LEARNING ACTIVITY 8.1

State how incineration helps in the management of solid wastes.

Note:

a) Write your answer in the space given below.

b) Check your answer with the one given at the end of this book.

8.2 PLANNING AN INCINERATION FACILITY

Incineration of solid wastes is becoming an increasingly important aspect of solid waste management, as communities look for alternatives to rapidly filling landfills (or disappearing landfill sites). Modern incineration facilities are no longer simple garbage burners. Instead, they are designed to produce steam and electricity and can be used as a complement to source reduction, recycling and composting programmes. However, strategic long-term planning is essential for developing a successful incineration facility. In other words, it is important to develop an understanding of a variety of issues in the planning process, including the following:

(i) **Facility ownership and operation:** One of the first planning decisions that local officials make is about the entity that will actually own the facility and oversee its operation. This decision is based largely on the amount of financial risk the community is willing to assume and the time and resources available. Some of the procurement options, in this context, are:

- **Full service approach:** In this system, the community specifies only the process type and the performance required, and hires a (single) firm to design, construct and operate the plant.
- **Merchant plants:** In this type of system, in which waste is accepted on weight basis, a private firm designs, constructs, owns and operates the facility.
- **Turnkey approach:** In this system, a single company designs and builds the plant, according to the communities' specifications, and the community or a different contractor owns and operates the plant.

(ii) **Energy market:** Waste incineration facilities differ from most government services in that they generate a product as well as energy, which are sold for revenue. Steam and electricity are the energy products at incineration facilities, depending on the particular design.

(iii) **Marketing steam:** The primary end uses of steam from waste incineration facilities are industrial and institutional heating and cooling systems. Marketing of steam products involves identifying these industries and institutions within the region. Industrial and institutional steam users include textile, paper and pulp, food processing, leather, chemical producers, hospitals, etc. Planning must include proper backup to guarantee a consistent supply and steam demand variation (often caused by changing seasons).

(iv) **Marketing electricity:** Incineration facilities generating electricity are referred to as *co-generators* as they provide electricity in addition to that generated by the local electric utility. Besides the plant that uses the electricity generated for its operation, customers for electricity include nearby industries and public, and private utilities. It must be equipped to give a consistent supply and must compete with other co-generators in selling energy.

(v) **Facility siting:** Siting the incineration facility is one of the most important tasks to be undertaken and a variety of social and technical hurdles have to be negotiated. The important aspects in this context are the following:

- **Effect on residents:** Residents will be concerned with the health effects associated with incinerator plant, decreased property value and increased traffic (e.g., due to truck movement).
- **Environmental impact:** Incineration has the potential to create a variety of environmental concerns like air, water and noise pollution and ash disposal.
- **Development plans:** It is important to evaluate future land use plans at the possible site.
- **Proximity to waste source:** Transportation cost is one of the most significant expenditures in waste management system.
- **Proximity to energy market:** The energy products will have to be delivered to buyers. The location of power line must be considered.
- **Logistic concerns:** Area zoning and access route must be considered.
- **Residual ash disposal:** Access to a secure landfill is necessary.

(vi) **Facility sizing:** Proper plant sizing results from careful evaluation of a wide variety of criteria such as:

- **Waste supply:** This is the most fundamental sizing factor and measures are usually taken to guarantee a waste supply for the facility. Waste flow control ordinances are often used to ascertain the quantity of waste. When properly planned, waste flow control can benefit both incineration facility and alternative waste management programme, by diverting the relevant portions of the waste stream (e.g., recyclables to the recycling programme and combustibles to incineration facility).
- **Alternative waste management programme:** In addition to waste flow control agreements, future source reduction, recycling and composting programmes are directly related to facility design. When sizing the incineration facility, it is important, therefore, to account for the type and amount of materials that will be diverted from the facility.
- **Waste stream characteristics:** Good combustion depends on the accuracy of waste stream data. Planning of incineration facility requires waste stream assessment to develop an accurate picture of the quantity and composition of the waste stream. From a technical standpoint, the waste stream data will be used to ascertain the heating value of the waste, which helps in plant operation.

- **Planning for facility disruption:** Accounting for downtime is an important facility planning criteria. Most incineration facilities are designed to operate continuously (i.e., 24 hours a day), but both scheduled (e.g., maintenance) and unscheduled (e.g., equipment failure) downtime situations are likely to occur. Storage space must be available for the waste that continues to arrive during downtime. If these capabilities are not built into the system, provisions must be made to send waste to a landfill or an alternative facility.
- **Facility financing:** Depending upon the procurement approach selected, incineration facility will require extensive financing agreements.
- **Time frame:** The time required to plan, develop and construct a facility will vary, but at least 5 to 8 years are required to bring a new facility from the early planning stages to in-service.

Long-term planning within the local government is the key for successful facility design and operation. By understanding all issues and dedicated workforce, waste combustion can become a positive component of waste management system (EPA, 1989 and 1995).

✍ LEARNING ACTIVITY 8.2

List the various planning stages of thermal treatment.

Note:

a) Write your answer in the space given below.

b) Check your answer with the one given at the end of this book.

In order to assess the technical feasibility of the use of an incinerator, we will have to look into the various technologies that are available. In Section 8.3, we will examine some of these technologies.

8.3 INCINERATION TECHNOLOGIES

The four incineration technologies covered in this section are mass burning system, refuse-derived fuel system, modular incineration and fluidised bed incineration. The two most widely used and technically proven incineration technologies are mass-burning incineration and modular incineration. Fluidised-bed incineration has been employed to a lesser extent, although its use has been expanding and experience with this relatively new technology has increased. Refuse-derived fuel production and incineration has also been used, with limited success. Some facilities have been used in conjunction with pyrolysis, gasification and other related processes that convert solid waste to gaseous, liquid, or solid fuel through thermal processing (UNEP, 1996). In Subsections 8.3.1 to 8.3.4, we will discuss the four incineration technologies.

8.3.1 Mass-burning System

Mass-burning systems are the predominant form of MSW incineration. A mass-burn facility typically consists of a reciprocating grate combustion system and a refractory-lined, water-walled steam generator. Mass-burn systems generally consist of either two or three incineration units ranging in capacity from 50 to 1,000 tonnes per day. That is to say, the facility capacity ranges from about 100–150 to 2,000–3,000 tonnes per day. These facilities can accept refuse that has undergone little preprocessing other than the removal of oversized items. Although this versatility makes mass-burn facilities convenient and flexible, local programmes to separate household hazardous wastes (e.g., cleaners and pesticides) and recover certain materials (e.g., iron scrap) are necessary to help ensure environmentally viable incineration and resource conservation.

Because of the larger facility size, an incineration unit is specially designed to efficiently combust the waste to recover greater quantities of steam or electricity for revenue. To achieve this greater combustion and heat recovery efficiency, the larger field-erected incinerators are usually in-line furnaces with a grate system. The steam generator generally consists of refractory-coated water wall systems, i.e., walls comprised of tubes through which water circulates to absorb the heat of combustion. In a water wall system, the boiler is an integral part of the system wall, rather than a separate unit as is in a refractory system.

Mass-burning of waste can also be achieved by the use of a rotary kiln. Rotary kilns use a turning cylinder, either refractor or water wall design, to tumble the waste through the system. The kiln is reclined, with waste entering at the high elevation end and ash and non-combustibles leaving at the lower end.

The waste intake area usually includes a tipping floor, pit, crane and sometimes conveyors. Trucks enter the tipping floor and tip their wastes either onto the floor itself, or directly into the pit. When wastes are tipped onto the floor, a front-end loader or a bulldozer is used to push them into the pit or onto a conveyor. From a feed chute, MSW is continuously fed to a grate system, which moves the waste through a combustion chamber using a tumbling motion. A travelling or reciprocating grate may follow rotary combustors to further complete combustion.

At least two combustor units are included to provide a level of redundancy and to allow waste processing at a reduced rate during periods of scheduled and unscheduled maintenance. Mass-burn facilities today generate a higher quality of steam (i.e., pressure and temperature), which is then passed through a turbine generator to produce electricity or through an extraction turbine to generate electricity as well as provide process steam for heating or other purposes. Figure 8.1 below illustrates a typical mass-burn facility.

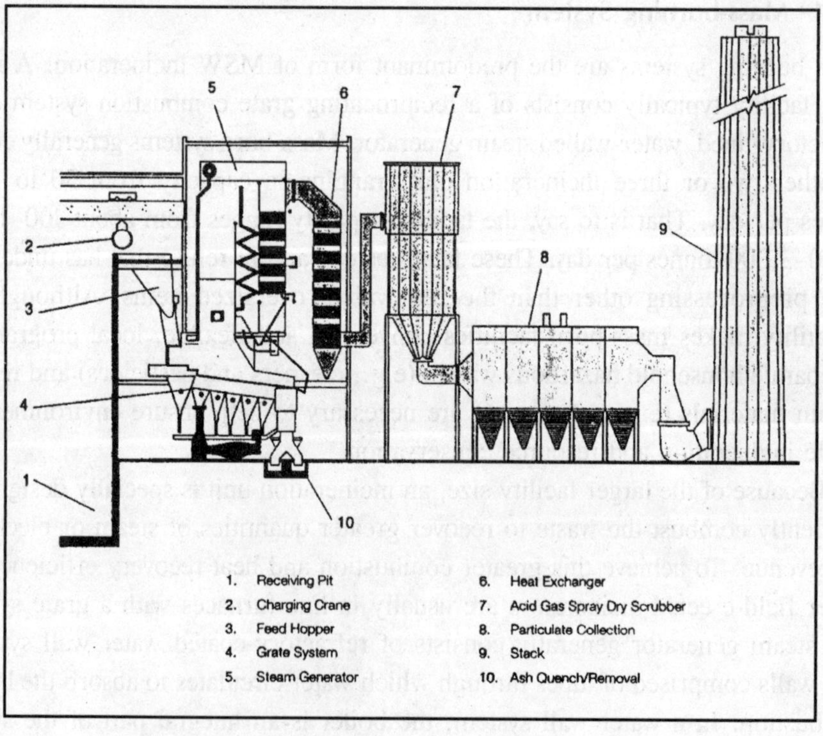

Figure 8.1 Typical Mass-Burn Facility

8.3.2 Refuse Derived Fuel (RDF) System

Refuse-derived fuel (RDF) refers to solid wastes in any form that is used as fuel. The term RDF, however, is commonly used to refer to solid waste that has been mechanically processed to produce a storable, transportable and more homogeneous fuel for combustion. RDF systems have two basic components: RDF production and RDF incineration.

RDF production facilities make RDF in various forms through material separation, size reduction and pelletising. Although RDF processing has the advantage of removing recyclables and contaminants from the combustion stream, on an average, capital costs per tonne for incineration units that use RDF are higher than for other incineration options. RDF production plants like mass-burn incinerators characteristically have an indoor tipping floor. Instead of being pushed onto a pit, the waste in an RDF plant is typically fed onto a conveyor, which is either below grade or hopper fed. In some plants, the loader doing the feeding will separate corrugated and bulky items like carpets.

Once on the conveyor, the waste travels through a number of processing stages, usually beginning with magnetic separation. The processing steps are tailored to the desired products, and typically include one or more screening stages, using trammel or vibrating screens, shredding or hammer milling of waste with additional screening steps, pelletising or baling of burnable wastes and, depending on the local recycling markets and the design of the facility, a manual separation line. A typical RDF facility scheme is given in Fig. 8.2 below.

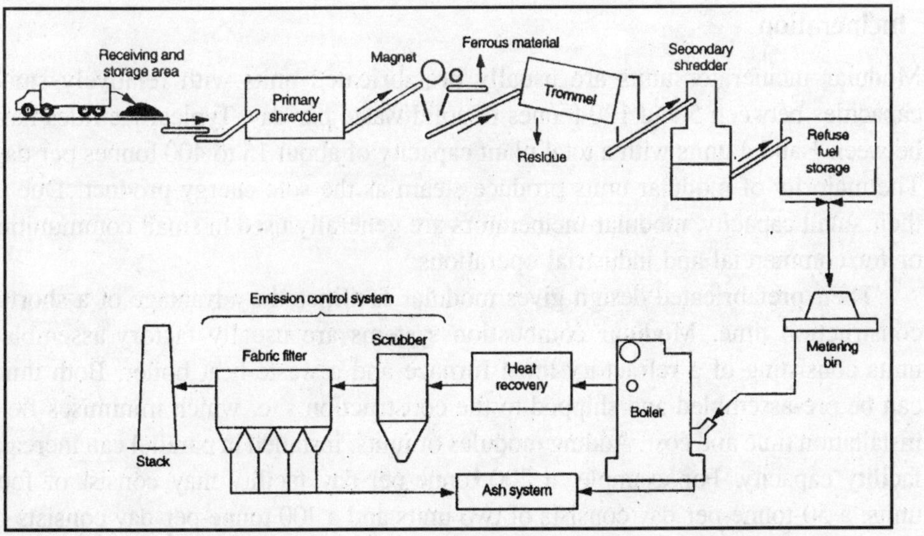

Figure 8.2 Typical Simplified RDF Facility

There are two primary types of systems in operation, and they are:

(i) **Shred-and-burn systems:** Shred-and-burn systems are the simplest form of RDF production. The process system typically consists of shredding the MSW to the desired particle size that allows effective feeding to the combustor and magnetic removal of ferrous metal, with the remaining portion delivered to the combustor. There is no attempt to remove other non-combustible materials in the MSW before combustion. This, in essence, is a system with minimal processing and removal of non-combustibles.

(ii) **Simplified process systems:** This is a system that removes a significant portion of the non-combustibles. A simplified process system involves processing the MSW to produce an RDF with a significant portion of the non-combustibles removed before combustion. The MSW process removes more than 85% of the ferrous metals, a significant percentage of the remaining non-combustible (i.e., glass, nonferrous metals, dirt, sand, etc.), and shreds the material to a nominal particle top size of 10 to 15 cm to allow effective firing in the combustion unit.

Depending on the type of combustor to be used, a significant degree of separation can be achieved to produce a high-quality RDF (i.e., low ash), which typically results in the loss of a higher percentage of combustibles when compared to systems that can produce a low-quality fuel (i.e., slightly higher ash content) for firing in a specially designed combustor. These types of systems recover over 95% of the combustibles in the fuel fraction.

8.3.3 Modular Incineration

Modular incinerator units are usually prefabricated units with relatively small capacities between 5 and 120 tonnes of solid waste per day. Typical facilities have between 1 and 4 units with a total plant capacity of about 15 to 400 tonnes per day. The majority of modular units produce steam as the sole energy product. Due to their small capacity, modular incinerators are generally used in small communities or for commercial and industrial operations.

Their prefabricated design gives modular facilities the advantage of a shorter construction time. Modular combustion systems are usually factory-assembled units consisting of a refractory-lined furnace and a waste heat boiler. Both units can be pre-assembled and shipped to the construction site, which minimises field installation time and cost. Adding modules or units, installed in parallel can increase facility capacity. For example, a 200 tonne-per-day facility may consist of four units, a 50-tonne-per-day consists of two units and a 100 tonne-per-day consists of one unit. The number of units may depend on the fluctuation of waste generation for the service area and the anticipated maintenance cycle of the units.

Modular incinerators employ a different process from that of mass-burn incinerators, typically involving two combustion chambers, and combustion is typically achieved in two stages.

The first stage may be operated in a condition in which there is less than the theoretical amount of air necessary for complete combustion. The controlled air condition creates volatile gases, which are fed into the secondary chamber, mixed with additional combustion air, and under controlled conditions, completely burned. Combustion temperatures in the secondary chamber are regulated by controlling the air supply, and when necessary, through the use of an auxiliary fuel. The hot combustion gases then pass through a waste heat boiler to produce steam for electrical generation or for heating purposes. The combustion gases and products are processed through air emission control equipment to meet the required emission standards.

In general, modular incineration systems are a suitable alternative and may, for smaller-sized facilities, be more cost-effective than other incinerators. But modular incineration has become less common, partly due to concerns over the consistency and adequacy of air pollution controls.

8.3.4 Fluidised-bed Incineration

Fluidised-bed incineration of MSW is typically medium scale, with processing capacity from 50 to 150 tonnes per day. In this system, a bed of limestone or sand that can withstand high temperatures, fed by an air distribution system, replaces the grate. The heating of the bed and an increase in the air velocities cause the bed to bubble, which gives rise to the term *fluidised*. There are two types of fluidised-bed technologies, viz., bubbling bed and circulating bed. The differences are reflected in the relationship between air flow and bed material, and have implications for the type of wastes that can be burned, as well as the heat transfer to the energy recovery system.

Unlike mass-burn incinerators, fluidised-bed incinerators require front-end pre-processing, also called fuel preparation. They are generally associated with source separation because glass and metals do not fare well in these systems and also they can successfully burn wastes of widely varying moisture and heat content, so that the inclusion of paper and wood, which are both recyclable and burnable, is not a crucial factor in their operation (and thus paper can be extracted for higher-value recycling). Fluidised-bed systems are more consistent in their operation than mass burn and can be controlled more effectively to achieve higher energy conversion efficiency, less residual ash and lower air emissions. In general, however, these systems appear to operate efficiently on smaller scales than mass-burn incinerators, which may make them attractive in some situations.

For this reason, fluidised-bed technology may be a sound choice for high-recycling cities in developing countries when they first adopt incineration.

✍ LEARNING ACTIVITY 8.3

State the advantages and disadvantages of mass-burning system.

Note:

a) Write your answer in the space given below.

b) Check your answer with the one given at the end of this book.

The benefit of incineration is a substantial reduction in the weight and volume of waste, and generation of revenue from energy production known as "waste-to-energy" (WTE), which can partially offset the cost of incineration. Keeping this in view, we will discuss the various options of energy generation from waste.

8.4 ENERGY RECOVERY

Most of the MSW incineration currently practice energy recovery in the form of steam, which is used either to drive a turbine to generate electricity or directly for heating or cooling. In the past, it was common to simply burn MSW in incinerators to reduce its volume and weight, but now energy recovery has become more prevalent (EPA, 1989 and 1995).

In waste-to-energy (WTE) plants, heat from the burning waste is absorbed by water in the wall of the furnace chamber, or in separate boilers. Water when heated to the boiling point changes to steam. At this point, the steam is used either for heating or to turn turbines to generate electricity. The amount of energy recovered from waste is a function of the amount of waste combusted, energy value of the waste stream and the efficiency of the combustion process (UNEP, 1996). The three basic types of waste-to-energy incineration are:

(i) **Generation of electricity:** Electricity is the most common form of energy produced and sold from WTE facilities constructed today. By directing the steam produced from a WTE system through a turbine generator, electricity can be produced and sold. A process flow diagram of an electrical generation system is shown in Fig. 8.3.

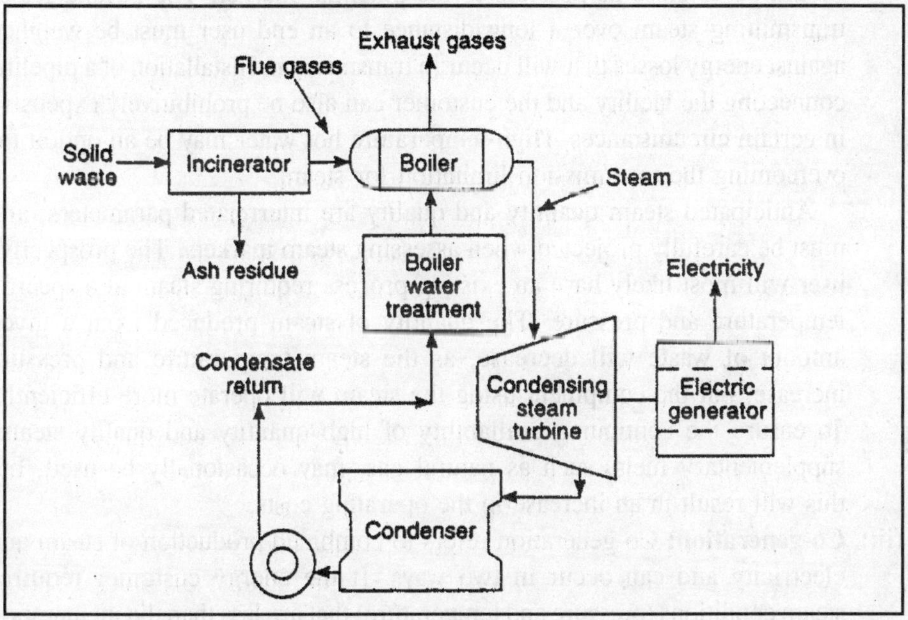

Figure 8.3 Incinerator and Electrical Generation System

Since electric utilities can receive power 24 hours a day and are usually very stable financially, public utilities have attractive markets for power produced from WTE systems. Of the electricity produced in incineration facilities, about one-fifth is used at the facility for general operations, and the remainder is sold to public and private utilities or nearby industries. In many countries, utilities provide a stable market for electricity generated from incinerators. The availability of purchasers and rates for electricity sales will, however, vary by region.

(ii) **Steam generation**: Steam is used widely in a variety of industrial applications. Steam generated in incineration facilities can also be used directly by a customer for manufacturing operations. Steam generated in an incinerator is supplied to a customer through a steam line, and a separate line sometimes returns the condensed steam. It can be used to drive machinery such as compressors, for space heating and generating electricity. Industrial plants, dairies, cheese plants, public utilities, paper mills, tanneries, breweries, public buildings and many other businesses use steam for heating and air conditioning.

When assessing potential markets for steam, it is important to consider a market's proximity to the WTE facility and the quantity of steam produced. Proximity is important because steam cannot usually be economically transported more than one or two miles. The WTE facility, therefore, should be as close as possible to the potential market. The advantages of transmitting steam over a long distance to an end user must be weighed against energy losses that will occur in transmission. Installation of a pipeline connecting the facility and the customer can also be prohibitively expensive in certain circumstances. High-temperature hot water may be an option for overcoming the transmission limitation for steam.

Anticipated steam quantity and quality are interrelated parameters, and must be carefully projected when assessing steam markets. The prospective user will most likely have an existing process requiring steam at a specific temperature and pressure. The quantity of steam produced from a given amount of waste will decrease, as the steam temperature and pressure increase, but the equipment using the steam will operate more efficiently. To ensure the continuing availability of high quantity and quality steam, supplementary fuels, such as natural gas, may occasionally be used, but this will result in an increase in the operating costs.

(iii) **Co-generation:** Co-generation refers to combined production of steam and electricity and can occur in two ways. If the energy customer requires steam conditions (pressure and temperature) that are less than the incineration plant's design specifications, a turbine-generator is used to produce electricity and thus reduce steam conditions to appropriate levels for the customer. If the steam purchaser cannot accept all the steam produced by the facility, the excess can be converted to electricity. In co-generation, high-pressure steam is used first to generate electricity; the steam leaving the turbine is then used to serve the steam users. Co-generation (Fig. 8.4) provides greater overall energy efficiency, even though the output of the major energy product – whether electricity or steam – may be less than that generated by producing one type of energy alone.

Co-generation allows flexibility, so that seasonal variations in steam demand can be offset by increase in electricity production, and can provide the project a financial base by selling electricity, should the steam customer become unavailable.

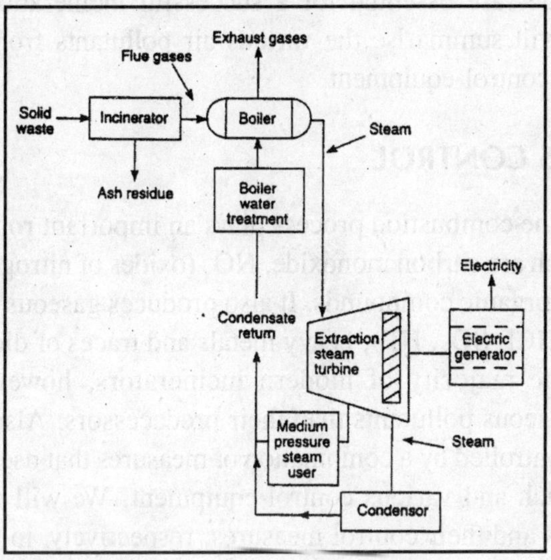

Figure 8.4 Co-generation System for Producing Electricity and Steam

✍ LEARNING ACTIVITY 8.4

Describe the three types of waste-to-energy incineration.

Note:

a) Write your answer in the space given below.
b) Check your answer with the one given at the end of this book.

The major risks of incinerators are the potential emission of contaminants into the air through exhaust stack or into the water through ash leachate. Proper planning to minimise these problems as well as public education and involvement that directly address these issues are essential for a successful incineration programme. In Section 8.5, we will summarise the various air pollutants from the incineration process and their control equipment.

8.5 AIR EMISSION AND ITS CONTROL

The operation of the combustion process plays an important role in the formation of pollutants, which are carbon monoxide, NO_x (oxides of nitrogen), hydrocarbons and other volatile organic compounds. It also produces gaseous stream containing dust, acid gases (HCl, SO_x, HF), heavy metals and traces of dioxins (McDougall et al., 2001). The majority of modern incinerators, however, produce less particulate and gaseous pollutants than their predecessors. Also, emissions from incinerators are controlled by a combination of measures that use both the pollution prevention approach and various control equipment. We will describe the main gaseous pollutants and their control measures, respectively, in Subsections 8.5.1 and 8.5.2.

8.5.1 Gaseous Pollutants

The various gaseous pollutants formed due to incineration processes are:

- **Carbon dioxide (CO_2):** This is one of the main products of incineration, and the other main product is water. At low concentrations, CO_2 has no short-term toxic or irritating effect, as it is abundant in the atmosphere and necessary for plant life and is not generally considered a pollutant. Nevertheless, due to the high increase in global concentration of CO_2, it has been recognised as one of the gases responsible for global warming.
- **Carbon monoxide (CO):** An incomplete combustion of carbon due to the lack of oxygen forms CO. This gas is toxic, as it reacts with the haemoglobin in the blood, causing a decrease of available oxygen to the organisms. This lack of oxygen produces headache, nausea, suffocation and eventually death. Carbon monoxide in the flue gas is used to monitor the incomplete combustion of the other emissions, such as un-burnt hydrocarbons and provide information on the performance of the incinerator.
- **Sulphur oxides (SO_x):** The emission of SO_x is a direct result of the oxidation of sulphur present in solid waste, but other conditions such as the type of incinerator used and its operating conditions also influence its production. Approximately 90% of SO_x emissions are SO_2 and 10% are SO_3. In the

atmosphere, most of the SO_2 is transformed into SO_3, which leads to the production of H_2SO_3 (sulphurous acid) and H_2SO_4 (sulphuric acid), increasing the acidity of rain. At high concentrations, it causes eye, nose and throat irritation, and other respiratory problems.

- **Nitrogen oxides (NO_x):** This is predominantly formed during the incineration process. However, they oxidise to NO_2 in the atmosphere. NO_x is formed from two main sources – thermal NO_x and fuel NO_x. In thermal formation, the oxygen and nitrogen react in the air. Fuel NO_x is formed during the reactions between oxygen and nitrogen in the fuel. Nitrogen oxides are important, as they participate in several processes in atmospheric chemistry. They are precursors of the formation of ozone (O_3) and peroxy acetal nitrate (PAN). These photochemical oxidants are responsible for smog formation and cause acid rain.
- **Particulates:** This is formed during the combustion process by several mechanisms. The turbulence in the combustion chambers may carry some ash into the exhaust flow. Other inorganic materials present in the waste volatilise at combustion temperature and later condense downstream to form particles or deposits on ash particles. The main component of fly ash is chemically inert silica; but it may also contain toxic metal and organic substances.
- **Hydrochloric acid (HCl):** Hydrochloric acid results from the high concentration of chlorine containing materials (e.g., some type of plastics like polyvinyl chloride) in solid waste. Chlorine easily dissolves in water to form HCl. Its presence in the gaseous state may increase the acidity of local rain and ground water, which can damage exposed and unprotected metal surfaces, erode buildings and may affect the mobilisation of heavy metals in soil.
- **Hydrogen fluoride (HF):** Hydrogen fluoride is more toxic and corrosive than HCl, although its presence in the emissions from solid waste incinerators occurs in much smaller quantities. It is formed due to the presence of trace amounts of fluorine in the waste.
- **Heavy metals (Hg, Cd, Pb, Zn, Cu, Ni, Cr):** Solid waste contains heavy metals and metallic compounds in the combustible and incombustible fractions. During the incineration process, metals may vapourise directly or form oxides or chlorides at high temperatures in the combustion zone. They condensate over other particles and leave the incineration process in the flue gas.
- **Dioxins and furans:** Polychlorinated dibenzo-p-dioxins (PCDD) and polychlorinated dibenzofurans (PCDF) have been detected in the emissions from solid waste incinerators. Dioxins can be formed in all combustion processes, where organic carbon, oxygen and chlorine are present, although the processes by which they are formed during incineration are not completely understood.

The concern over dioxins and furans has increased after a number of animal studies have shown that for some species, they are carcinogenic and highly toxic, even at very low levels of exposure.

8.5.2 Gas-cleaning Equipment

The technologies employed to carry out the necessary flue gas cleaning include:

- **Electrostatic precipitators (ESP):** These are used for particle control. ESP use electric forces to move the particle flowing out of the gas stream on to the collector electrodes. The particles get a negative charge when they pass through an ionised field. The electric field that forces the charged particles to the walls comes from discharged electrodes maintained at high voltage in the centre of the flow lane.

  When particles are collected, they must be carefully removed to avoid their entry into the gaseous stream. This is achieved by knocking them loose from the plates and by intermittent and continuous washing with water. The removal efficiency is more than 99% with a low-pressure drop. Particle size and other physical characteristics such as gas stream temperature, flue gas volume, moisture content, gas stream composition, particle composition and particle surface characteristics affect the performance of the equipment. ESP's come in one- or two-stage designs with different configurations: plate wire, flat plate and tubular. These configurations can be wet or dry, depending on the method of dust collection.
- **Fabric filters:** In fabric filtration, the gas flows through a number of filter bags placed in parallel, leaving the dust captured by the fabric. Extended operation of fabric filters requires periodic cleaning of the cloth surface. After a new fabric goes through a number of cycles of use and cleaning, it forms a residual cake of dust that becomes the filter medium, which is responsible for highly efficient filtering of small particles that characterises fabric filter. They are widely accepted for controlling particulate matter.

  The type of cloth fabric limits the temperature of operation of fabric cloth: cotton is least resistant (82 °C) while fabric glass is the most resistant (277 °C). This temperature requirement makes the use of a cooling system necessary for the gas, before it enters the equipment, but it is also necessary that the temperature of the exhaust gas stream be maintained above the dew point because liquid particles block the pores in the fabric very quickly. The major difference between different configurations of fabric filters is the cleaning method used during operating cycles such as shaker, reverse-air and pulse-jet cleaning.

- **Scrubbers:** Scrubbers are used to control particulate matter and acid gases leaving the incinerator. Particulates and gases are removed from the gas stream mainly by absorption and adsorption. The particles are moved to the vicinity of the water droplets and collide with each other. The particles adhere to the liquid media and precipitate to the bottom of the unit containing the dust particle from the gas phase. In addition to removing entrapped particulate matter, scrubbers can also remove gases by absorption and adsorption. Any other type of particulate control equipment does not possess this capability of scrubber. It can remove particles of size 0.1 to 200 μm efficiently. The various types of scrubbers can be dry, semi-dry or wet, depending on the composition of flue gas.

✍ LEARNING ACTIVITY 8.5

Describe the mechanism of an electrostatic precipitator.

Note:

a) Write your answer in the space given below.

b) Check your answer with the one given at the end of this book.

8.6 ENVIRONMENTAL CONCERNS

In Section 8.5, we discussed the various air pollutants emitted from an incinerator and their control equipment. Apart from air pollution, there are other environmental concerns related to incineration (EPA, 1989 and 1995), some of which are touched upon below:

(i) **Water pollution:** Wastewater in an incineration facility can be generated in various forms. These include tipping floor runoff system wash water, ash quench water and water from pollution control systems. These systems also deal with normal problems experienced by all large industrial facilities, including sanitary wastewater disposal and surface-water runoff. For most incineration facilities, wastewater can be recycled in a closed-loop system. In these systems, water from floor drains, ash dewatering, water softener recharge and other process wastewaters are collected and stored in a tank. This water is then reused for ash quenching. Sanitary waste can be directed to municipal sewer systems. In some cases, regulatory authorities may require that the waste stream be pretreated before discharge.

(ii) **Land-retained pollution:** Land-retained pollutants originate as stack or fugitive emissions and are of increasing concern. Bioaccumulation and subsequent ingestion from food is an indirect exposure route resulting from land-retained emissions. To provide better understanding of land-retained pollutants, it may be desirable to establish baseline contaminant levels, before incineration plant construction so that changes in those levels throughout the plant's operating lifetime can be monitored.

(iii) **Residue disposal:** An incineration facility and its emission control system produce a variety of residues such as large quantity of bottom ash, the unburned and not burnable materials discharged from the incinerator at the end of the burning cycle. The process also produces a lighter emission known as fly ash. Fly ash consists of products in particulate form, which are either produced as a result of the chemical decomposition of burnable materials or are unburned (or partially burned) materials drawn upward by thermal air currents in the incinerator and trapped in pollution control equipment. Fly ash normally comprises only a small proportion (ranging from 10 to 20%) of the total volume of residue from an incineration facility.

Constituents in both ash and scrubber product vary, depending on the materials burned. In systems burning a homogeneous fuel such as coal, oil, or tyres, the level of pollutants in residues may be relatively constant. Systems burning a more heterogeneous mixture, such as municipal, industrial, or medical waste may experience wide swings in the chemical composition of residues. The major constituents of concern in municipal waste combustion ash are heavy metals, particularly lead, cadmium and mercury. These metals may impact human health and the environment; if improperly handled, stored, transported, disposed off or reused.

(iv) **Noise pollution:** Truck traffic is the greatest source of noise pollution in incineration plant operations. Well maintained and responsibly operated trucks will help minimise this problem. Local ordinances may restrict truck traffic to certain hours of the day and to specified truck corridors. Under these conditions, noise pollution should not be a significant factor. Noise resulting from plant operations and air handling fans associated with the combustion and emission control equipment is also a potential problem. Noise levels are likely to be the highest in front of waste tipping floor doors, ash floor doors and the vicinity of air emission stacks. Walls, fences, trees and landscaped earthen barriers can serve to reduce noise levels.

(v) **Aesthetic impact:** Proper site landscaping and building design can help minimise or prevent negative aesthetic impacts. Such impacts are much less problematic, if the facility is sited in an industrial area and not adjacent to residential or commercial districts. Local zoning ordinances may ensure that aesthetic impacts are minimised. Keeping the process building at negative pressure can prevent undesirable odours from escaping outside of the building. Also, the use of internal air for combustion in the plant processes will destroy most odours.

Some facilities may emit visible steam or vapour plumes. Smoke, resulting from improper conditions in the combustion chamber, can also be problematic. Air emission stacks and cooling towers may also be unappealing anomalies in the skyline of some areas. If external lights on buildings prove objectionable to neighbours, perimeter lights (on stands) directed towards the plant may be preferable.

SUMMARY

Incineration is a chemical reaction by which carbon, hydrogen and other elements in the waste combine with oxygen in the combustible air and generate heat. In this Unit, we discussed the objectives and issues to be considered while planning an incineration facility. In this context, we discussed the various incineration processes and technologies such as mass burn, refuse derive fuel, modular incineration and fluidised bed incineration. We said that the product of incineration is heat that can be utilised for the generation of steam and electricity. We also pointed out that the amount of energy recovered is a function of the amount of waste combusted, the energy value of the waste stream and the efficiency of the combustion process. We then discussed energy generation. We also discussed the various air pollutants emitted out of an incineration facility and the equipment used for their control namely scrubbers, electrostatic precipitators and fabric filters. Finally, we listed some of the impacts of incinerators on the environment.

SUGGESTED READINGS

Ali, M., Cotton, A. and Westlake, K. 1999. *Down to Earth*. Loughborough University, UK.

Arne Vesilind, William Worrel and Reinhart Debra. 2002. *Solid Waste Engineering*. Thomson Brooks/Cole, Singapore.

Areivala, S.J. 1971. *Solid Wastes Disposal in India*. Central Public Health Engineering Institute, Nagpur.

Attarwalla, F.A. 1993. *Solid Waste Management - A Manual*. All India Institute of Local Self-Government, Andheri (W), Bombay, India.

Environmental Management in Developing Countries. 1995. *Waste Management Vol II*. Institute for Scientific Co-operation, Tubingen.

Flinotoff, R. 1984. *Management of Solid Wastes in Developing Countries*. WHO Regional Publications, Southeast Asia Series No 1.

Ramachandra, T.V. and Saira Varghese, K. 2004. Environmentally sound options for e-wastes management. *ENVIS Journal of Human Settlements* (ISSN 0971-9091), March 2004.

Ramachandra, T.V. and Saira Varghese, K. 2003. Exploring possibilities of achieving sustainability in solid waste management. *Indian J. Environmental Health*, **45 (4):** 255-264.

Ogawa, H. 1989. Selection of Appropriate Technology for Solid Waste Management in Asian Metropolises, *An International Journal of Regional Development Dialogue*. UNCRD, Nagoya, Japan, Vol. 10, No. 3, Autumn.

Tchobanoglous, G., Theisen, H. and Eliassan, R. 1977. *Solid Wastes—Engineering Principles and Management Issues*. McGraw-Hill Book Company, New York.

Tchobanoglous, G., Theisen, H. and Samuel A. Vigil. 1993. *Integrated Solid Waste Management*. McGraw-Hill, Inc., New Delhi.

UNEP. 1996. *International Source Book on Environmentally Sound Technologies for Municipal Solid Waste Management (6)*. IETC, Osaka/Shiga.

US Environmental Protection Agency. 1989. *Decision-Maker's Guide to Solid Waste Management, Vol I*. Washington.

US Environmental Protection Agency. 1995. *Decision-Maker's Guide to Solid Waste Management, Vol II*. Washington.

http://ces.iisc.ernet.in/energy/SWMTR/TR85.html

REFERENCES

McDougall, F.R., White, P.R., Franke, M. and Hindle, P. 2001. *Integrated Solid Waste Management: A Life Cycle Inventory*. Blackwell Science, UK.

Phelps, H.O., Heinke, G.W., Jonker, J.F., Ouano, E.A.R. and Vandecasteele, C. 1995. *Management of Solid Wastes*. UNESCO, Paris.

Tchobanoglous, G., Theisen, H. and Eliassan, R. 1977. *Solid Wastes—Engineering Principles and Management Issues*. McGraw-Hill Book Company, New York.

UNEP. 1996. *International Source Book on Environmentally Sound Technologies for Municipal Solid Waste Management (6),* IETC, Osaka/Shiga.

US Environmental Protection Agency. 1989. *Decision-Maker's Guide to Solid Waste Management, Vol I*. Washington.

US Environmental Protection Agency. 1995. *Decision-Maker's Guide to Solid Waste Management, Vol II*. Washington.

UNIT 9

Hazardous Waste: Management and Treatment

STRUCTURE

OVERVIEW

In Units 1 to 8, we discussed the management of solid waste and its functional elements, which include storage, collection, transport, waste disposal, processing, recycling, biological conversion of waste and incineration with energy recovery in the context of non-hazardous wastes. In view of the substantial threat – present and potential – hazardous wastes pose to human health, or living organisms in general, they ought to be handled, treated and managed differently, and this issue is discussed in the present Unit. In this Unit, we will first identify and classify hazardous wastes, and then discuss their functional elements namely generation, storage and collection, transfer and transport, processing and disposal.

Subsequently, we will discuss the various physical, chemical, thermal and biological treatments to reduce the impact of hazardous wastes on public health and the environment. We will close the Unit by explaining some of the techniques for hazardous waste minimisation and pollution prevention and touching upon the prevailing hazardous waste management practices in India.

LEARNING OBJECTIVES

After completing this Unit, you should be able to:

- identify and classify hazardous wastes;
- explain the techniques of hazardous waste management, treatment and minimisation;
- describe the physical, chemical, thermal and biological methods of treating hazardous waste;
- adopt waste minimisation and pollution prevention techniques.

9.1 HAZARDOUS WASTE: IDENTIFICATION AND CLASSIFICATION

Hazardous wastes refer to wastes that may, or tend to, cause adverse health effects on the ecosystem and human beings. These wastes pose present or potential risks to human health or living organisms, duc to the fact that they:

- are non-degradable or persistent in nature;
- can be biologically magnified;
- are highly toxic and even lethal at very low concentrations.

The above list relates only to the intrinsic hazard of the waste, under uncontrolled release, to the environment, regardless of quantity or pathways to humans or other critical organisms (i.e., plants and animals). The criteria used to determine the nature of hazard include toxicity, phytotoxicity, genetic activity and bio-concentration. The threat to public health and the environment of a given hazardous waste is dependent on the quantity and characteristics of the waste involved. Wastes are secondary materials, which are generally classified into six categories as inherently waste: like materials, spent materials, sludges, by-products, commercial chemical products and scrap metals. Solid wastes form a subset of all secondary materials and hazardous wastes form a subset of solid waste. However, note that certain secondary materials are not regulated as wastes, as they are recycled and reused.

Figure 9.1 illustrates the relationship among secondary materials, solid wastes and hazardous wastes *(http://www.dep.state.pa.us/dep/deputate/airwaste/)*.

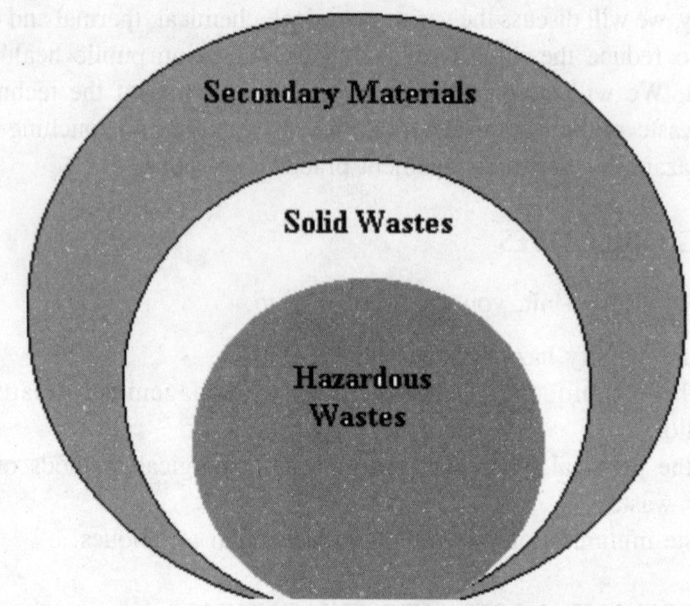

Figure 9.1 Secondary Materials, Solid and Hazardous Wastes: Relationship

Note that for a material to be classified as a hazardous waste, it must also meet the criteria specified in the regulatory definition of solid waste, which we will study next.

9.1.1 Identification

By using either or both of the following criteria, we can identify as to whether or not a waste is hazardous:

(i) The list provided by government agencies declaring that substance as hazardous.
(ii) Characteristics such as ignitability, corrosivity, reactivity and toxicity of the substance.

Let us now explain these two criteria.

Listed hazardous wastes (priority chemicals)

A specific list showing certain materials as hazardous wastes minimises the need to test wastes as well as simplifies waste determination. In other words, any waste that fits the definition of a listed waste is considered a hazardous waste. Four separate lists cover wastes from generic industrial processes, specific industrial sectors, unused pure chemical products and formulations that are either acutely toxic or toxic, and all hazardous waste regulations apply to these lists of wastes.

We will describe these wastes, classified in the F, K, P, and U industrial waste codes, respectively, below *(http://www2.kumc.edu/safety/kdhehw/hwg1_10.html#SectionIV)*:

- **F-list:** The F-list contains hazardous wastes from non-specific sources, that is, various industrial processes that may have generated the waste. The list consists of solvents commonly used in degreasing, metal treatment baths and sludges, wastewaters from metal plating operations and dioxin containing chemicals or their precursors. Examples of solvents that are F-listed hazardous wastes, along with their code numbers, include benzene (F005), carbon tetrachloride (F001), cresylic acid (F004), methyl ethyl ketone (F005), methylene chloride (F001), 1,1,1, trichloroethane (F001), toluene (F005) and trichloroethylene (F001). Solvent mixtures or blends, which contain greater than 10% of one or more of the solvents listed in F001, F002, F003, F004 and F005 are also considered F-listed wastes.
- **K-list:** The K-list contains hazardous wastes generated by specific industrial processes. Examples of industries, which generate K-listed wastes include wood preservation, pigment production, chemical production, petroleum refining, iron and steel production, explosive manufacturing and pesticide production.
- **P and U lists:** The P and U lists contain discarded commercial chemical products, off-specification chemicals, container residues and residues from the spillage of materials. These two lists include commercial pure grades of the chemical, any technical grades of the chemical that are produced or marketed, and all formulations in which the chemical is the sole active ingredient. An example of a P or U listed hazardous waste is a pesticide, which is not used during its shelf-life and requires to be disposed in bulk. The primary distinction between the two lists is the quantity at which the chemical is regulated. The P-list consists of acutely toxic wastes that are regulated when the quantity generated per month, or accumulated at any time, exceeds one kilogram (2.2 pounds), while U-listed hazardous wastes are regulated when the quantity generated per month exceeds 25 kilograms (55 pounds). Examples of businesses that typically generate P or U listed wastes include pesticide applicators, laboratories and chemical formulators.

Characteristics of hazardous wastes

The regulations define characteristic hazardous wastes as wastes that exhibit measurable properties posing sufficient threats to warrant regulation. For a waste to be deemed a characteristic hazardous waste, it must cause, or significantly contribute to, an increased mortality or an increase in serious irreversible or incapacitating reversible illness, or pose a substantial hazard or threat of a hazard to human health or the environment, when it is improperly treated, stored, transported, disposed off, or otherwise mismanaged.

In other words, if the wastes generated at a facility are not listed in the F, K, P, or U lists, the final step to determine whether a waste is hazardous is to evaluate it against the following four hazardous characteristics:

(i) **Ignitability** (EPA Waste Identification Number D001): A waste is an ignitable hazardous waste, if it has a flash point of less than 60 °C; readily catches fire and burns so vigorously as to create a hazard; or is an ignitable compressed gas or an oxidiser. A simple method of determining the flash point of a waste is to review the material safety data sheet, which can be obtained from the manufacturer or distributor of the material. Naphtha, lacquer thinner, epoxy resins, adhesives and oil-based paints are all examples of ignitable hazardous wastes.

(ii) **Corrosivity** (EPA Waste Identification Number D002): A liquid waste which has a pH of less than or equal to 2 or greater than or equal to 12.5 is considered to be a corrosive hazardous waste. Sodium hydroxide, a caustic solution with a high pH, is often used by many industries to clean or degrease metal parts. Hydrochloric acid, a solution with a low pH, is used by many industries to clean metal parts prior to painting. When these caustic or acid solutions are disposed off, the waste is a corrosive hazardous waste.

(iii) **Reactivity** (EPA Waste Identification Number D003): A material is considered a reactive hazardous waste, if it is unstable, reacts violently with water, generates toxic gases when exposed to water or corrosive materials, or if it is capable of detonation or explosion when exposed to heat or a flame. Examples of reactive wastes would be waste gunpowder, sodium metal or wastes containing cyanides or sulphides.

(iv) **Toxicity** (EPA Waste Identification Number D004): To determine if a waste is a toxic hazardous waste, a representative sample of the material must be subjected to a test conducted in a certified laboratory. The toxic characteristic identifies wastes that are likely to leach dangerous concentrations of toxic chemicals into ground water.

9.1.2 Classification

From a practical standpoint, there are far too many compounds, products and product combinations that fit within the broad definition of hazardous waste. For this reason, groups of waste are considered in the following five general categories:

(i) **Radioactive substance:** Substances that emit ionising radiation are radioactive. Such substances are hazardous because prolonged exposure to radiation often results in damage to living organisms. Radioactive substances are of special concern because they persist for a long period. The period in which radiation occurs is commonly measured and expressed as *half-life*, i.e., the time required for the radioactivity of a given amount of the substance to decay to half its initial value. For example, uranium compounds have half-lives that range from 72 years for U_{232} to 23,420,000 years for U_{236}. The management of radioactive wastes is highly controlled by national and state regulatory agencies. Disposal sites that are used for the long-term storage of radioactive wastes are not used for the disposal of any other solid waste.

(ii) **Chemicals:** Most hazardous chemical wastes can be classified into four groups: synthetic organics, inorganic metals, salts, acids and bases, and flammables and explosives. Some of the chemicals are hazardous because they are highly toxic to most life forms. When such hazardous compounds are present in a waste stream at levels equal to, or greater than, their threshold levels, the entire waste stream is identified as hazardous.

(iii) **Biomedical wastes:** The principal sources of hazardous biological wastes are hospitals and biological research facilities. The ability to infect other living organisms and the ability to produce toxins are the most significant characteristics of hazardous biological wastes. This group mainly includes malignant tissues discarded during surgical procedures and contaminated materials, such as hypodermic needles, bandages and outdated drugs. This waste can also be generated as a by-product of industrial biological conversion processes.

(iv) **Flammable wastes:** Most flammable wastes are also identified as hazardous chemical wastes. This dual grouping is necessary because of the high potential hazard in storing, collecting and disposing off flammable wastes. These wastes may be liquid, gaseous or solid, but most often they are liquids. Typical examples include organic solvents, oils, plasticisers and organic sludges.

(v) **Explosives:** Explosive hazardous wastes are mainly ordnance (artillery) materials, i.e., the wastes resulting from ordnance manufacturing and some industrial gases. Similar to flammables, these wastes also have a high potential for hazard in storage, collection and disposal, and therefore, they should be considered separately in addition to being listed as hazardous chemicals. These wastes may exist in solid, liquid or gaseous form.

LEARNING ACTIVITY 9.1

Identify the characteristics of a major hazardous waste generated in your locality.

Note:

a) Write your answer in the space given below.

b) Check your answer with the one given at the end of this book.

We will now discuss the management of hazardous waste, which is different from the management of other solid wastes due to hazardous nature of wastes.

9.2 HAZARDOUS WASTE MANAGEMENT

Hazardous waste management, as is the case with non-hazardous solid waste management, which we studied earlier, consists of several functional elements. We will discuss these elements in Subsections 9.2.1 to 9.2.5.

9.2.1 Generation

Hazardous wastes are generated in limited amounts in a community and very little information is available on the quantities of hazardous waste generated within a community and in various industries. Hazardous waste generation outside the industry is irregular and very less in amount, rendering the waste generation parameter meaningless. The only practical means to overcome these limitations is to conduct a detailed inventory and measurement studies at each potential source in a community. As a first step in developing a community inventory, potential sources of hazardous waste are to be identified. The total annual quantity of hazardous waste at any given source in a community must be established through data inventory completed during onsite visits.

Table 9.1 below presents a list of hazardous waste generation sources.

Table 9.1 Common Hazardous Wastes: Community Source

Waste Category	Sources
Radioactive substances	Biomedical research facilities, colleges and university laboratories, offices, hospitals, nuclear power plants, etc.
Toxic chemicals	Agricultural chemical companies, battery shops, car washes, chemical shops, college and university laboratories, construction companies, electric utilities, hospitals and clinics, industrial cooling towers, newspaper and photographic solutions, nuclear power plants, pest control agencies, photographic processing facilities, plating shops, service stations, etc.
Biological wastes	Biomedical research facilities, drug companies, hospitals, medical clinics, etc.
Flammable wastes	Dry cleaners, petroleum reclamation plants, petroleum refining and processing facilities, service stations, tanker truck cleaning stations, etc.
Explosives	Construction companies, dry cleaners, ammunition production facilities, etc.

Source: Tchobanoglous et al. (1977 and 1993)

In addition to the sources listed, the spillage of containerised hazardous waste must also be considered an important source. The quantities of hazardous wastes that are involved in spillage are usually not known. The effects of spillage are often spectacular and visible to the community. Because the occurrence of spillage cannot be predicted, the potential threat to human health and environment is greater than that from routinely generated hazardous wastes.

9.2.2 Storage and Collection

Onsite storage practices are a function of the types and amounts of hazardous wastes generated and the period over which generation occurs. Usually, when large quantities are generated, special facilities are used that have sufficient capacity to hold wastes accumulated over a period of several days. When only a small amount is generated, the waste can be containerised, and limited quantity may be stored. Containers and facilities used in hazardous waste storage and handling are selected on the basis of waste characteristics. For example, corrosive acids or caustic solutions are stored in fibreglass or glass-lined containers to prevent deterioration of metals in the container. Great care must also be exercised to avoid storing incompatible wastes in the same container or locations. Figures 9.2 and 9.3 show typical drum containers used for the storage of hazardous waste.

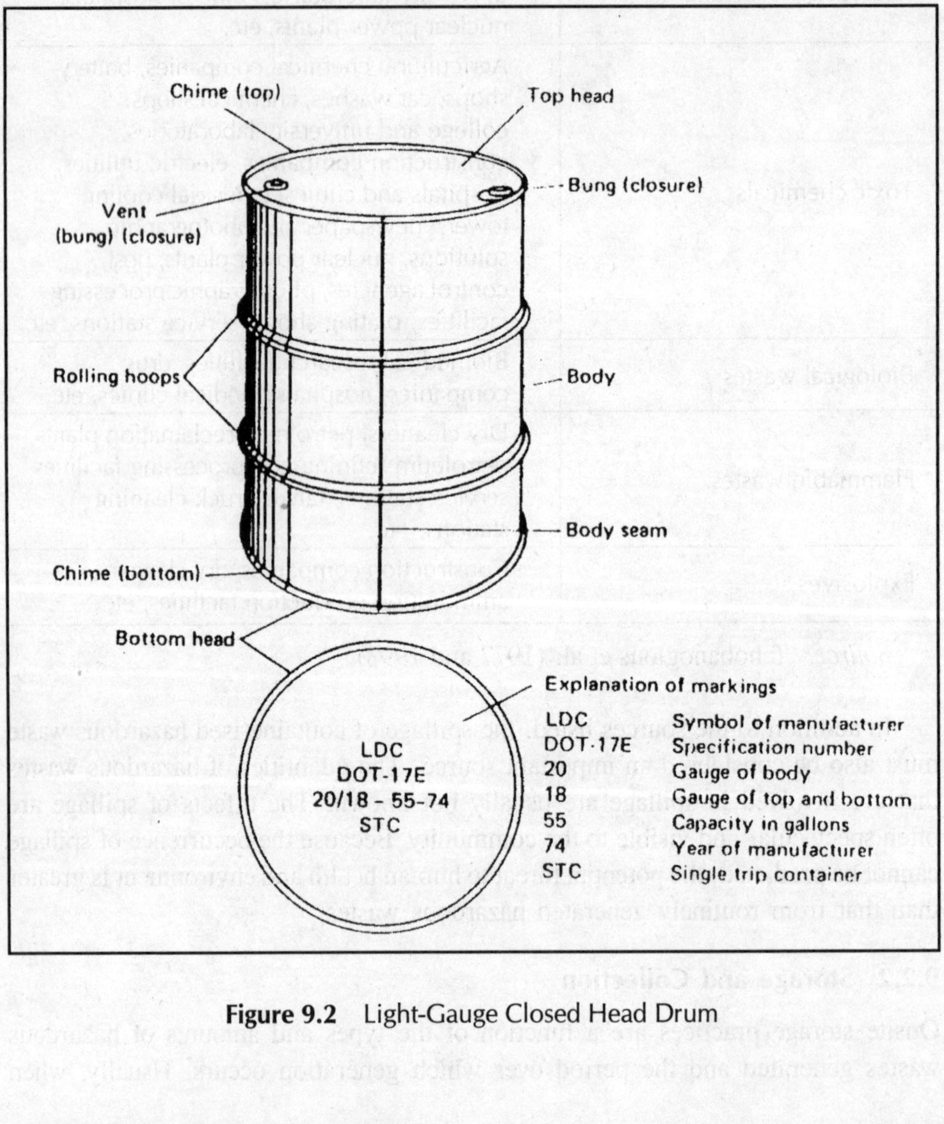

Figure 9.2 Light-Gauge Closed Head Drum

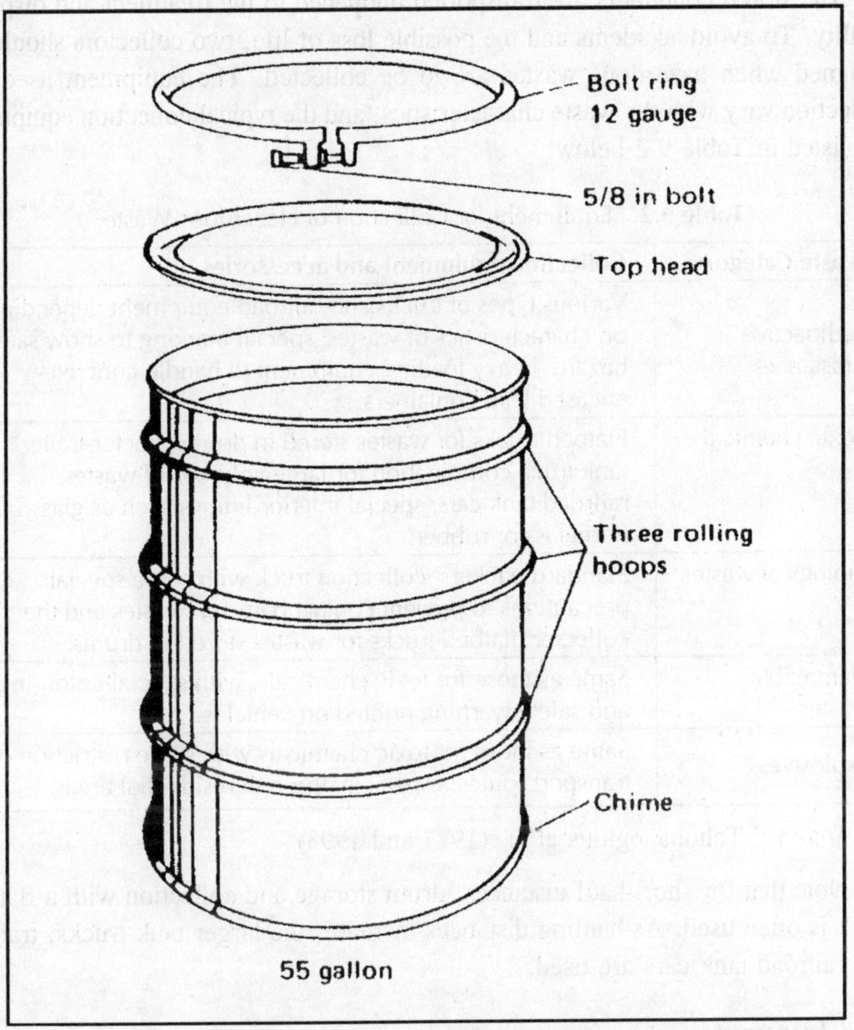

Figure 9.3 Light-Gauge Open Head Drum

The waste generator, or a specialised hauler, generally collects the hazardous waste for delivery to a treatment or disposal site. The loading of collection vehicles is completed in either of the following ways:

(i) Wastes stored in large-capacity tanks are either drained or pumped into collection vehicles;

(ii) Wastes stored in sealed drums or sealed containers are loaded by hand or by mechanical equipment onto flatbed trucks.

The stored containers are transported unopened to the treatment and disposal facility. To avoid accidents and the possible loss of life, two collectors should be assigned when hazardous wastes are to be collected. The equipment used for collection vary with the waste characteristics, and the typical collection equipment are listed in Table 9.2 below.

Table 9.2 Equipment for Collection of Hazardous Waste

Waste Category	Collection equipment and accessories
Radioactive substances	Various types of trucks and railroad equipment depending on characteristics of wastes; special marking to show safety hazard; heavy loading equipment to handle concrete-encased lead containers.
Toxic chemicals	Flatbed trucks for wastes stored in drums; tractor-trailer tank truck combination for large volumes of wastes; railroad tank cars; special interior linings such as glass, fibreglass or rubber.
Biological wastes	Standard packers' collection truck with some special precautions to prevent contact between wastes and the collector; flatbed trucks for wastes stored in drums.
Flammable wastes	Same as those for toxic chemicals, with special colourings and safety warning printed on vehicles.
Explosives	Same as those for toxic chemicals with some restriction on transport routes, especially through residential areas.

Source: Tchobanoglous et al. (1977 and 1993)

Note that for short-haul distances, drum storage and collection with a flatbed truck is often used. As hauling distances increase, the larger tank trucks, trailers and railroad tank cars are used.

9.2.3 Transfer and Transport

The economic benefits derived by transferring smaller vehicle loads to larger vehicles, as discussed for non-hazardous solid waste in Unit 3, are equally applicable to hazardous wastes. However, the facilities of a hazardous waste transfer station are quite different from solid waste transfer station. Typically, hazardous wastes are not compacted (i.e., mechanical volume reduction) or delivered by numerous community residents. Instead, liquid hazardous wastes are generally pumped from collection vehicles and sludge or solids are reloaded without removal from the collection containers for transport to processing and disposal facilities.

It is unusual to find a hazardous waste transfer facility, where wastes are simply transferred to larger transport vehicles. Some processing and storage facilities are often part of the material handling sequence at a transfer station. For example, neutralisation of corrosive wastes might result in the use of a lower-cost holding tank on transport vehicles. As in the case of storage (as discussed above in this subsection), great care must be exercised to avoid the danger of mixing incompatible wastes.

9.2.4 Processing

Processing of hazardous waste is done for purposes of recovering useful materials and preparing the wastes for disposal.

Processing can be accomplished on-site or off-site. The variables affecting the selection of processing site include the characteristics of wastes, the quantity of wastes, the technical, economical and environmental aspects of available on-site treatment processes and the availability of the nearest off-site treatment facility (e.g., haul distance, fees and exclusions). The treatment of hazardous waste can be accomplished by physical, chemical, thermal or biological means. Table 9.3 below gives the various individual processes in each category.

Table 9.3 Hazardous Waste Treatment Operations and Processes

Operation/Processes	Functions performed$	Types of wastes*	Forms of waste#
Physical Treatment			
Aeration	Se	1, 2, 3, 4	L
Ammonia stripping	VR, Se	1, 2, 3, 4	L
Carbon sorption	VR, Se	1, 3, 4, 5	L,G
Centrifugation	VR, Se	1, 2, 3, 4, 5	L
Dialysis	VR, Se	1, 2, 3, 4	L
Distillation	VR, Se	1, 2, 3, 4, 5	L
Electro dialysis	VR, Se	1, 2, 3, 4, 6	L
Encapsulation	St	1, 2, 3, 4, 6	L,S
Evaporation	VR, Se	1, 2, 5	L
Filtration	VR, Se	1, 2, 3, 4, 5	L,G
Flocculation/Settling	VR, Se	1, 2, 3, 4, 5	L
Flotation	Se	1, 2, 3, 4	L
Reverse osmosis	VR, Se	1, 2, 4, 6	L
Sedimentation	VR, Se	1, 2, 3, 4, 5	L
Thickening	Se	1, 2, 3, 4	L
Vapour scrubbing	VR, Se	1, 2, 3, 4	L
Chemical Treatment			
Calcination	VR	1, 2, 5	L

(Contd.)

Table 9.3 (*Contd.*)

Ion exchange	VR, Se, De	1, 2, 3, 4, 5	L
Neutralisation	De	1, 2, 3, 4	L
Oxidation	De	1, 2, 3, 4	L
Precipitation	VR, Se	1, 2, 3, 4, 5	L
Reduction	De	1, 2	L
Solvent extraction	Se	1, 2, 3, 4, 5	L
Sorption	De	1, 2, 3, 4	L
Thermal treatment			
Incineration	VR, De	3, 5, 6, 7, 8	S, L, G
Pyrolysis	VR, De	3, 4, 6	S, L, G
Biological Treatment			
Activated sludges	De	3	L
Aerated lagoons	De	3	L
Anaerobic digestion	De	3	L
Anaerobic filters	De	3	L
Trickling filters	De	3	L
Waste stabilisation pond	De	3	L

Source: Tchobanoglous et al. (1977, 1993)

$ *Functions*: VR= volume reduction; Se = separation; De = detoxification; St = storage;
* *Waste types*: 1= inorganic chemical without heavy metals; 2 = inorganic chemical with heavy metal; 3 = organic chemical without heavy metal; 4 = organic chemical with heavy metal; 5= radiological; 6 = biological; 7= flammable and 8= explosive; # *Waste forms*: S=solid; L= liquid and G= gas

Note that in practice, the physical, chemical and thermal treatment operations are the most commonly used. (Biological treatment processes are used less often because of their sensitivity.) Depending on the type of wastes being treated, one or more of these methods may be used. We will explain some of these methods later in Section 9.3.

9.2.5 Disposal

Regardless of their form (i.e., solid, liquid or gas), most hazardous waste is disposed off either near the surface or by deep burial. Table 9.4 shows the various hazardous waste disposal methods.

Table 9.4 Hazardous Wastes Disposal and Storage Methods

Operation/Processes	Functions performed[$]	Types of wastes[*]	Forms of waste[#]
Deep well injection	Di	1, 2, 3, 4,5,6,7	L
Detonation	Di	6, 8	S, L, G
Engineered storage	St	1, 2, 3, 4, 5,6,7,8	S, L, G
Land burial	Di	1, 2, 3, 4, 5,6,7,8	S, L
Ocean dumping	Di	1, 2, 3, 4, 7, 8	S, L, G

Source: Tchobanoglous et al. (1977 and 1993)

[$] *Functions*: Di= disposal; St = storage; [*] *Waste types*: 1= inorganic chemical without heavy metals; 2 = inorganic chemical with heavy metal; 3 = organic chemical without heavy metal; 4 = organic chemical with heavy metal; 5= radiological; 6 = biological; 7= flammable and 8= explosive. [#] *Waste form*: S=solid; L= liquid and G= gas

Although, controlled landfill methods have been proved adequate for disposing off municipal solid waste and limited amounts of hazardous waste, they are not suitable enough for the disposal of a large quantity of hazardous waste, due to the following reasons:

- possible percolation of toxic liquid waste to the ground water;
- dissolution of solids followed by leaching and percolation to the ground water;
- dissolution of solid hazardous wastes by acid leachate from solid waste, followed by leaching and percolation to the ground water;
- potential for undesirable reactions in the landfill that may lead to the development of explosive or toxic gases;
- volatilisation of hazardous waste leading to the release of toxic or explosive vapours to the atmosphere;
- corrosion of containers with hazardous wastes.

We must, therefore, take care both in the selection of a hazardous waste disposal site and its design. In general, disposal sites for hazardous wastes should be separate from those for municipal solid wastes. As hazardous wastes can exist in the form of liquids, sludges, solids and dusts, a correct approach for co-disposal for each of the hazardous wastes should be determined. To avoid the co-disposal of incompatible wastes, separate storage areas within the total landfill site should be designated for various classes of compatible wastes (Phelps et al., 1995).

Liquid wastes are usually stored in a tank near the site and can be introduced into the landfill by means of trenches or lagoons, injection or irrigation. Sludges are also placed in trenches. During disposal of lightweight wastes, the disposal area must be kept wet to prevent dust emissions. Hazardous solid waste characterised by a high degree of impermeability as such must not be disposed off over large areas. When containerised wastes are to be disposed off, precautions must be taken to avoid the rupturing of containers during the unloading operation and the placement of incompatible waste in the same location. To avoid rupturing, the containers are unloaded and placed in position individually. The covering of the containers with earth should be monitored and controlled carefully to ensure that a soil layer exists between each container and the equipment placing the soil does not crush or deform the container.

While designing a landfill site for hazardous waste, provision should be made to prevent any leachate escaping from landfill site. This requires a clay liner, and in some cases, both clay and impermeable membrane liners are used. A layer of limestone is placed at the bottom of the landfill to neutralise the pH of leachate. A final soil cover of 25 cm or more should be placed over the liner. The completed site should be monitored continuously, both visually and with sample wells.

✍ LEARNING ACTIVITY 9.2

Suggest the ways and means of properly managing hazardous wastes.

Note:

a) Write your answer in the space given below.

b) Check your answer with the one given at the end of this book.

9.3 HAZARDOUS WASTE TREATMENT

In Section 9.2, we discussed the various elements of hazardous waste management such as generation, storage and transport, transfer and transport, processing and disposal. Processing is mainly done to recover useful products and to prepare waste for disposal. But prior to disposal, hazardous wastes need appropriate treatment, depending on the type of waste. The various options for hazardous waste treatment can be categorised under physical, chemical, thermal and biological treatments. We will discuss these options, in Subsections 9.3.1 to 9.3.3.

9.3.1 Physical and Chemical Treatment

Physical and chemical treatments are an essential part of most hazardous waste treatment operations, and the treatments include the following (Freeman, 1988):

(i) **Filtration and separation:** Filtration is a method for separating solid particles from a liquid using a porous medium. The driving force in filtration is a pressure gradient, caused by gravity, centrifugal force, vacuum, or pressure greater than atmospheric pressure. The application of filtration for treatment of hazardous waste fall into the following categories:

- **Clarification,** in which suspended solid particles less than 100 ppm (parts per million) concentration are removed from an aqueous stream. This is usually accomplished by depth filtration and cross-flow filtration and the primary aim is to produce a clear aqueous effluent, which can either be discharged directly, or further processed. The suspended solids are concentrated in a reject stream.
- **Dewatering** of slurries of typically 1% to 30 % solids by weight. Here, the aim is to concentrate the solids into a phase or solid form for disposal or further treatment. This is usually accomplished by cake filtration. The filtration treatment, for example, can be used for neutralisation of strong acid with lime or limestone, or precipitation of dissolved heavy metals as carbonates or sulphides followed by settling and thickening of the resulting precipitated solids as slurry. The slurry can be dewatered by cake filtration and the effluent from the settling step can be filtered by depth filtration prior to discharge.

(ii) **Chemical precipitation:** This is a process by which the soluble substance is converted to an insoluble form either by a chemical reaction or by change in the composition of the solvent to diminish the solubility of the substance in it. Settling and/or filtration can then remove the precipitated solids. In the treatment of hazardous waste, the process has a wide applicability in the removal of toxic metal from aqueous wastes by converting them to an insoluble form. This includes wastes containing arsenic, barium, cadmium, chromium, copper, lead, mercury, nickel, selenium, silver, thallium and zinc. The sources of wastes containing metals are metal plating and polishing, inorganic pigment, mining and the electronic industries. Hazardous wastes containing metals are also generated from cleanup of uncontrolled hazardous waste sites, e.g., leachate or contaminated ground water.

(iii) **Chemical oxidation and reduction (redox):** In these reactions, the oxidation state of one reactant is raised, while that of the other reactant is lowered. When electrons are removed from an ion, atom, or molecule, the substance is oxidised and when electrons are added to a substance, it is reduced. Such reactions are used in treatment of metal-bearing wastes, sulphides, cyanides and chromium and in the treatment of many organic wastes such as phenols, pesticides and sulphur containing compounds. Since these treatment processes involve chemical reactions, both reactants are generally in solution. However, in some cases, a solution reacts with a slightly soluble solid or gas.

There are many chemicals which are oxidising agents but relatively few of them are used for waste treatment. Some of the commonly used oxidising agents are sodium hypochlorite, hydrogen peroxide, calcium hypochlorite, potassium permanganate and ozone. Reducing agents are used to treat wastes containing hexavalent chromium, mercury, organometallic compounds and chelated metals. Some of the compounds used as reducing agents are sulphur dioxide, sodium borohydride, etc. In general, chemical treatment costs are highly influenced by the chemical cost. This oxidation and reduction treatment tends to be more suitable for low concentration (i.e., less than 1%) in wastes.

(iv) **Solidification and stabilisation:** In hazardous waste management, solidification and stabilisation (S/S) is a term normally used to designate a technology employing activities to reduce the mobility of pollutants, thereby making the waste acceptable under current land disposal requirements. Solidification and stabilisation are treatment processes designed to improve waste handling and physical characteristics, decrease surface area across which pollutants can transfer or leach, limit the solubility or detoxify the hazardous constituent. To understand this technology, it is important for us to understand the following terms:

- **Solidification:** This refers to a process in which materials are added to the waste to produce a solid. It may or may not involve a chemical bonding between the toxic contaminant and the additive.
- **Stabilisation:** This refers to a process by which a waste is converted to a more chemically stable form. Subsuming solidification, stabilisation represents the

use of a chemical reaction to transform the toxic component to a new, non-toxic compound or substance.

- **Chemical fixation:** This implies the transformation of toxic contaminants to a new non-toxic compound. The term has been misused to describe processes, which do not involve chemical bonding of the contaminant to the binder.
- **Encapsulation:** This is a process involving the complete coating or enclosure of a toxic particle or waste agglomerate with a new substance (e.g., S/S additive or binder). The encapsulation of the individual particles is known as micro-encapsulation, while that of an agglomeration of waste particles or micro-encapsulated materials is known as macro-encapsulation.

In S/S method, some wastes can be mixed with filling and binding agents to obtain a dischargeable product. This rather simple treatment can only be used for waste with chemical properties suitable for landfilling. With regard to wastes with physical properties, it changes only the physical properties, but is unsuitable for landfilling. The most important application of this technology, however, is the solidification of metal-containing waste. S/S technology could potentially be an important alternative technology with a major use being to treat wastes in order to make them acceptable for land disposal. Lower permeability, lower contaminant leaching rate and such similar characteristics may make hazardous wastes acceptable for land disposal after stabilisation.

(v) **Evaporation:** Evaporation is defined as the conversion of a liquid from a solution or slurry into vapour. All evaporation systems require the transfer of sufficient heat from a heating medium to the process fluid to vaporise the volatile solvent. Evaporation is used in the treatment of hazardous waste and the process equipment is quite flexible and can handle waste in various forms – aqueous, slurries, sludges and tars. Evaporation is commonly used as a pre-treatment method to decrease quantities of material for final treatment. It is also used in cases where no other treatment method was found to be practical, such as in the concentration of trinitrotoluene (TNT) for subsequent incineration.

(vi) **Ozonation:** Ozone is a relatively unstable gas consisting of three oxygen atoms per molecule (O_3) and is one of the strongest oxidising agents known. It can be substituted for conventional oxidants such as chlorine, hydrogen peroxide and potassium permanganate. Ozone and UV radiations have been used to detoxify industrial organic wastes, containing aromatic and aliphatic polychlorinated compounds, ketones and alcohols.

9.3.2 Thermal Treatment

The two main thermal treatments used with regard to hazardous wastes are:

(i) **Incineration:** Incineration can be regarded as either a pre-treatment of hazardous waste, prior to final disposal or as a means of valorising waste by recovering energy. It includes both the burning of mixed solid waste or burning of selected parts of the waste stream as a fuel. The concept of treating hazardous waste is similar to that of municipal solid waste (see Unit 8).

(ii) **Pyrolysis:** This is defined as the chemical decomposition or change brought about by heating in the absence of oxygen. This is a thermal process for transformation of solid and liquid carbonaceous materials into gaseous components and the solid residue containing fixed carbon and ash. The application of pyrolysis to hazardous waste treatment leads to a two-step process for disposal. In the first step, wastes are heated separating the volatile contents (e.g., combustible gases, water vapour, etc.) from non-volatile char and ash. In the second step volatile components are burned under proper conditions to assure incineration of all hazardous components (Freeman, 1988).

To elaborate, pyrolysis is applicable to hazardous waste treatment, as it provides a precise control of the combustion process. The first step of pyrolysis treatment is endothermic and generally done at 425 to 760 °C. The heating chamber is called the pyrolyser. Hazardous organic compounds can be volatilised at this low temperature, leaving a clean residue. In the second step, the volatiles are burned in a fume incinerator to achieve destruction efficiency of more than 99%. Separating the process into two very controllable steps allows precise temperature control and makes it possible to build simpler equipment. The pyrolysis process can be applied to solids, sludges and liquid wastes. Wastes with the following characteristics are especially amenable to pyrolysis:

- Sludge material that is either too viscous, too abrasive or varies too much in consistency to be atomised in an incinerator.
- Wastes such as plastic, which undergo partial or complete phase changes during thermal processing.
- High-residue materials such as high-ash liquid and sludges, with light, easily entrained solids that will generally require substantial stack gas clean up.
- Materials containing salts and metals, which melt and volatilise at normal incineration temperatures. Materials like sodium chloride (NaCl), zinc (Zn) and lead (Pb), when incinerated may cause refractory spalling and fouling of the heat-exchanger surface.

9.3.3 Biological Treatment

On the basis of the fact that hazardous materials are toxic to living beings, it is not uncommon for some to assume that biological treatment is not possible for hazardous wastes. This assumption is untenable, and, in fact, we must aggressively seek biological treatment in order to exploit the full potential of hazardous wastes in terms of removal efficiency and cost (Freeman, 1988). Against this background, let us now list some of the techniques used for biological treatment of hazardous waste:

(i) **Land treatment:** This is a waste treatment and disposal process, where a waste is mixed with or incorporated into the surface soil and is degraded, transformed or immobilised through proper management. The other terminologies used commonly include land cultivation, land farming, land application and sludge spreading. Compared to other land disposal options (e.g., landfill and surface impoundments), land treatment has lower long-term monitoring, maintenance and potential clean up liabilities and because of this, it has received considerable attention as an ultimate disposal method. It is a dynamic, management-intensive process involving waste, site, soil, climate and biological activity as a system to degrade and immobilise waste constituents.

In land treatment, the organic fraction must be biodegradable at reasonable rates to minimise environmental problems associated with migration of hazardous waste constituents. The various factors involved in the operation of the system are as follows:

- **Waste characteristics:** Biodegradable wastes are suitable for land treatment. Radioactive wastes, highly volatile, reactive, flammable liquids and inorganic wastes such as heavy metals, acids and bases, cyanides and ammonia are not considered for land treatment. Land treatability of organic compound often follows a predictable pattern for similar type of compounds. Chemical structure, molecular weight, water solubility and vapour pressure are few of the characteristics that determine the ease of biodegradation.
- **Soil characteristics:** The rate of biodegradation and leaching of waste applied, the availability of nutrients and toxicants to microorganisms and the fate of hazardous waste constituents are determined largely by application rate as well as the soil's chemical and physical characteristics or reaction. Principal soil characteristics affecting land treatment processes are pH, salinity, aeration, moisture holding capacity, soil temperature, etc. Some of the characteristics can be improved through soil amendments (e.g., nutrients, lime, etc.), tillage or through adjustments of loading rate, frequency, etc., at the time of waste application.

- **Microorganisms:** Soil normally contains a large number of diverse microorganisms, consisting of several groups that are predominantly aerobic in well-drained soil. The types and population of microorganisms present in the waste-amended soil depend on the soil moisture content, available oxygen, nutrient composition and other characteristics. The key groups of the microorganisms present in the surface soil are bacteria, actinomycetes, fungi, algae and protozoa. In addition to these groups, other micro and macro fauna, such as nematodes and insects are often present.
- **Waste degradation:** Conditions favourable for plant growth are also favourable for the activity of soil microorganisms. The factors affecting waste degradation (that may be adjusted in the design and operation of a land treatment facility) are soil pH (near 7), soil moisture content (usually between 30 and 90 %), soil temperature (activity decreases below 10°C) and nutrients.

(ii) **Enzymatic systems:** Enzymes are complex proteins ubiquitous in nature. These proteins, composed of amino acids, are linked together via peptide bonds. Enzymes capable of transforming hazardous waste chemicals to non-toxic products can be harvested from microorganisms grown in mass culture. Such crude enzyme extracts derived from microorganisms have been shown to convert pesticides into less toxic and persistent products. The reaction of detoxifying enzymes are not limited to intracellular conditions but have been demonstrated through the use of immobilised enzyme extracts on several liquid waste streams. The factors of moisture, temperature, aeration, soil structure, organic matter content, seasonal variation and the availability of soil nutrients influence the presence and abundance of enzymes.

(iii) **Composting:** The principles involved in composting organic hazardous wastes are the same as those in the composting of all organic materials (See Unit 8), though with moderate modifications. The microbiology of hazardous wastes differs from that of composting in the use of inoculums. The reaction is that certain types of hazardous waste molecules can be degraded by only one or a very few microbial species, which may not be widely distributed or abundant in nature. The factors important in composting of hazardous wastes are those that govern all biological reactions. The principal physical parameters are the shape and dimensions of the particles of the material to be composted and the environmental factors of interest in an operation are temperature, pH, available oxygen, moisture and nutrient availability.

As we studied in Unit 7, the compost technology can be divided into two broad classes – windrow (open pile) and in-vessel (enclosed), and the former may be further subdivided into turned and forced aeration (static pile). Composting, by no means, is a panacea for the hazardous waste problem. When considering the future of hazardous waste composting needs, attention must be paid to the advantages and disadvantages inherent in composting as compared to those inherent in physical, chemical and thermal method of waste treatment.

(iv) **Aerobic and anaerobic treatment:** Hazardous materials are present in low to high concentration in wastewaters, leachate and soil. These wastes are characterised by high organic content (e.g., up to 40,000 mg/l total organic carbon), low and high pH (2 to 12), elevated salt levels (sometimes, over 5%), and presence of heavy metals and hazardous organics. Hazardous wastes can be treated using either aerobic or anaerobic treatment methods.

In aerobic treatment, under proper conditions, microorganisms grow. They need a carbon and energy source, which many hazardous wastes satisfy, nutrients such as nitrogen, phosphorus and trace metals and a source of oxygen. Some organisms can use oxidised inorganic compounds (e.g. nitrate) as a substitute for oxygen. Care is to be taken such that all the required nutrients and substances are supplied in sufficient quantities. Temperature and pH must be controlled as needed and the substances that are toxic to the organisms (e.g., heavy metals) must be removed.

Anaerobic treatment is a sequential biologically destructive process in which hydrocarbons are converted, in the absence of free oxygen, from complex to simpler molecules, and ultimately to carbon dioxide and methane. The process is mediated through enzyme catalysis and depends on maintaining a balance of population within a specific set of environmental conditions. Hazardous waste streams often consist of hydrocarbons leading to higher concentrations of chemical oxygen demand (COD). Depending upon the nature of waste, the organic constituents may be derived from a single process stream or from a mixture of streams.

The treatability of the waste depends upon the susceptibility of the hydrocarbon content to anaerobic biological degradation, and on the ability of the organisms to resist detrimental effect of biologically recalcitrant and toxic organic and inorganic chemicals. The metabolic interactions among the various groups of organisms are essential for the successful and complete mineralisation of the organic molecules. Various parameters such as the influent quality, the biological activity of the reactor and the quality of the reactor environment are monitored to maintain efficient operating conditions within the reactor.

LEARNING ACTIVITY 9.3

Explain the factors required for land treatment operation of hazardous wastes.

Note:

a) Write your answer in the space given below.

b) Check your answer with the one given at the end of this book.

9.4 POLLUTION PREVENTION AND WASTE MINIMISATION

Pollution prevention is the use of materials, processes, or practices that reduce or eliminate the generation of pollutants or wastes at the source. It includes practices that reduce the use of hazardous and non-hazardous materials, energy, water or other resources as well as those that protect natural resources through conservation or more efficient use. Pollution prevention is the maximum feasible reduction of all wastes generated at production sites. It involves the judicious use of resources through source reduction, energy efficiency, reuse of input materials and reduces water consumption.

Waste minimisation means the feasible reduction of hazardous waste that is generated prior to treatment, storage and disposal. It is defined as any source reduction or recycling activity that results in the reduction of the total volume of hazardous waste, or toxicity of hazardous waste, or both. Practices that are considered in waste minimisation include recycling, source separation, product substitution, manufacturing process changes and the use of less toxic raw materials.

Pollution prevention and waste minimisation provides us with an opportunity to be environmentally responsible (*http://www.ehs.umaryland.edu/waste/pollutio.htm*). While pollution prevention reduces waste at its source, waste minimisation, including recycling and other methods, reduces the amount of waste. In what follows, we will look at some of the factors that can contribute to pollution prevention and waste minimisation.

(i) **Management support and employee participation:** A clear commitment by management (through policy, communications and resources) for waste minimisation and pollution prevention is essential to earn the dedication of all employees. For this to happen, a formal policy statement must be drafted and adopted. The purpose of this statement is to reflect commitment and attitude towards protecting the environment, minimising or eliminating waste and reusing or recycling materials by the laboratories, departments and industries. Creative, progressive and responsible leadership will serve to develop an environmental policy. However, the total employee workforce will need to be involved to realise the fruits of the planning.

(ii) **Training:** As with any activity, it is important for management to train employees so that they will have an understanding of what is expected of them and why they are being asked to change the way things are done. Employees must be provided with formal and on-the-job training to increase awareness of operating practices that reduce both solid and hazardous waste generation. The training programme should include the industries' compliance requirements, which may be found in the waste management policies, occupational health and safety requirements. Additionally, training on waste minimisation and pollution prevention is necessary.

(iii) **Waste audits:** A programme of waste audits at the departmental level will provide a systematic and periodic survey of the industries designed to identify areas of potential waste reduction. The audit programme includes the identification of hazardous wastes and their sources, prioritisation of various waste reduction actions to be undertaken, evaluation of some technically, economically and ecologically feasible approaches to waste minimisation and pollution prevention, development of an economic comparison of waste minimisation and pollution prevention options and evaluation of their results.

(iv) **Good operating practices:** These practices involve the procedural or organisational aspects of industry, research or teaching activities and, in some areas, changes in operating practices, in order to reduce the amount of waste generated. These practices would include, at a minimum, material handling improvements, scheduling improvements, spill and leak prevention, preventive maintenance, corrective maintenance, material/waste tracking or inventory control and waste stream segregation, according to the toxicity, type of contaminant and physical state.

(v) **Material substitution practices:** The purpose of these practices is to find substitute materials, which are less hazardous than those currently utilised and which result in the generation of waste in smaller quantities and/or of less toxicity.

(vi) **Technological modification practices:** These practices should be oriented towards process and equipment modifications to reduce waste generation. These can range from changes that can be implemented in a matter of days at low cost to the replacement of process equipment involving large capital expenditures.

(vii) **Recycling options:** These options are characterised as use/reuse and resource recovery techniques. Use and reuse practices involve the return of a waste material either to the originating process or to another process as a substitute for an input material. Reclamation practices tender a waste to another company.

(viii) **Surplus chemical waste exchange options:** Inter- and intra-department chemical exchange is to be implemented and encouraged by employers/ employees. Material exchanges not only reduce wastes but also save money – both are important considerations, during times of fiscal crisis.

In addition, by auditing each department or section, a knowledge base of chemical purchase and usage can be developed, allowing each department to develop and implement controls on the purchase of chemicals, institute intra-departmental chemical sharing/swapping programmes and eliminate excessive purchase and usage.

Research protocols should also be examined and modified in a manner similar to the above. Facility operations need to be examined to determine whether changes in practices and procedures will result in the generation of non-hazardous or less hazardous waste, or waste reduced in toxicity or volume. The specifics to be considered in this context include the substitution of non-toxic materials for toxic ones, distillation or evaporation of water-based chemical end-products, reclamation and reuse of common solvents, use of non-chromate cleaners as a standard part of doing business to generate non-hazardous end products. By implementing and adhering to the guidelines for handling and storing wastes at the point of generation, the costs associated with hazardous waste disposal will also be minimised.

LEARNING ACTIVITY 9.4

Explain the advantages of waste minimisation and pollution prevention.

Note:

a) Write your answer in the space given below.

b) Check your answer with the one given at the end of this book.

9.5 HAZARDOUS WASTES MANAGEMENT IN INDIA

In the USA, more than 70% of the hazardous waste generated was produced from chemical and petrochemical industries. Of the remaining waste produced, 22% was generated by metal related industries. As industrialisation proceeds, the management of hazardous wastes is increasingly becoming a serious problem in India as well. The Indian chemical industry, which accounts for about 13% of the total industrial production and about 10% of the GNP valued at US \$ 2.64×10^{11} (NNP is US \$ 2.345×10^{11}) per annum, employs about 6% of the nation's industrial workforce and is one of the major generators of toxic and hazardous wastes. There are 13,011 industrial units located in 340 districts, out of which 11,038 units have been granted authorization for multiple disposal practices encompassing incineration, storage land disposal and other disposal options.

However, small and medium sized enterprises (SMEs) are the major sources of hazardous wastes. The States of Andhra Pradesh, Assam, Gujarat, Karnataka, Kerala, Madhya Pradesh, Maharashtra, Orissa, Rajasthan and Tamil Nadu generate the majority of all hazardous wastes. The total estimate of hazardous waste generated in India is 4,434,257 tonnes per annum.

India is the first country that has made provisions for the protection and improvement of environment in its Constitution. The Directive Principles of State Policy of the Constitution, Article 48A of Chapter IV enjoins the State to endeavour for protection and improvement of the environment and for safeguarding the forest and wild life of the country. In Article 51 A (g) of the Constitution, one of the Fundamental Duties of every citizen in India is to protect and improve the natural environment including forests, lakes, rivers and wild life and to have compassion for living creatures. India has enacted the following laws, regulations and standards governing the country's environmental protection:

(i) The Water (Prevention and Control of Pollution) Act, 1974 as amended in 1988.
(ii) Water (Prevention and Control of Pollution) Rules, 1975.
(iii) The Water (Prevention and Control of Pollution) Cess Act, 1977, as amended by Amendment Act, 1991.
(iv) The Water (Prevention and Control of Pollution) Cess Rules, 1978.
(v) The Air (Prevention and Control of Pollution) Act, 1984, as amended by Amendment Act, 1987.
(vi) The Air (Prevention and Control of Pollution) Rules 1982 and 1983.
(vii) The Environment (Protection) Act, 1986.
(viii) Hazardous Waste (Management and Handling) Rules, 1989 as amended in 2000.
(ix) Management, Storage and Import of Hazardous Chemical Rules, 1989.
(x) Manufacture, Use, Import, Export and Storage of Hazardous Microorganisms, Genetically Engineered Microorganisms or Cells Rules, 1989.
(xi) The Public Liability Insurance Act, 1991.
(xii) The Public Liability Insurance Rules, 1991.
(xiii) The Biomedical Wastes (Management and Handling) Rules, 1995.
(xiv) Municipal Wastes (Management and Handling) Draft Rules, 1999.
(xv) Hazardous Waste (Management and Handling) Amendment Rules, 2000.

Because of these amendments, the legal management of hazardous substances in India will now apply to 44 industrial processes, as specified in Schedule I of the Rules. The penal provisions for non-compliance under Hazardous Waste

(Management and Handling) Amended Rules 2000 and Environment (Protection) Act, 1986 are:

- The State Pollution Control Board may cancel an authorisation issued under these rules or suspend it for such period as it thinks fit, if, in its opinion, the authorised person has failed to comply with any of the conditions of the authorisation or with any provisions of the Act of these rules, after giving the authorised person an opportunity to show cause and after recording reasons therefore.
- The occupier, transporter and operator of a facility shall be liable for damages caused to the environment resulting due to improper disposal of hazardous waste listed in Schedule 1, 2 and 3 of The Hazardous Waste (Management and Handling) Amendment Rules, 2000. The occupier and operator of a facility shall also be liable to reinstate or restore damaged or destroyed elements of the environment. The occupier and operator of a facility shall be liable to pay a fine as levied by the SPCB with the approval of the Central Pollution Control Board (CPCB) for any violation of the provisions under these rules. An appeal shall lie against any order of grant or refusal of an authorisation by the Member Secretary, SPCB, etc., to the Secretary, Department of Environment of the State.

Besides the aforementioned provisions for non-compliance(s), the Penalty Provisions, delineated under Sections 15 (1, 2) and 16 of the Environmental (Protection) Act, 1986 are also applicable. Furthermore, the Union Ministry of Environment and Forests, through the Gazette Notification of March 24, 1992, introduced Public Liability Insurance Act Policy, which is specially designed to protect any person, firm, association, or company who owns or has control over handling any hazardous substance at the time of accident. These include 179 hazardous substances along with three categories of inflammable substances. The term *handling* means manufacturing, processing, treatment, packaging, storing, transportation by vehicle, use, collection, destruction, conversion, offering for sale, transfer or any other similar form of dealing with hazardous substances.

SUMMARY

In this Unit, our focus was on the management of hazardous wastes. We began the Unit with a discussion on the identification and classification of hazardous wastes, and in that context, explained the major identification lists (i.e., F, K, P and U), the characteristics (i.e., corrosivity, ignitability, toxicity and reactivity) and classifications (i.e., radioactive substances, chemicals, biological wastes, flammable wastes and explosives). We then discussed the functional elements involved in hazardous waste management (i.e., generation, storage and collection, transfer and transport, processing and disposal). Subsequently, we explained how hazardous wastes could be treated through the physical, chemical, thermal and biological means to reduce their impact on public health and the environment. We then discussed some of the techniques for hazardous waste minimisation and pollution prevention. We closed the Unit by touching upon some aspects of hazardous waste management in India.

SUGGESTED READINGS

Areivala, S.J. 1971. *Solid Wastes Disposal in India*. Central Public Health Engineering Institute, Nagpur.

Asian Productivity Council. 2001. Hazardous Waste Management: Policies and Practices in Asian Countries. Asian Productivity Council, Tokyo.

Arne Vesilind, William Worrel and Reinhart Debra. 2002. *Solid Waste Engineering*. Thomson Brooks/Cole, Singapore.

Tchobanoglous, G., Theisen, H. and Samuel A. Vigil. 1993. *Integrated Solid Waste Management*. McGraw-Hill, Inc., New Delhi.

Environmental Management in Developing Countries. 1995. *Waste Management Vol II*. Institute for Scientific Co-operation, Tubingen.

Flinotoff, R.1984. *Management of Solid Wastes in Developing Countries*. WHO Regional Publications, Southeast Asia Series No 1.

Freeman, M.H. 1988. *Standard Handbook of Hazardous Waste Treatment and Disposal*. McGraw-Hill Book Company, New York.

Ramachandra, T.V. and Saira Varghese, K. 2004. *Environmentally sound options for e-wastes management*. *ENVIS Journal of Human Settlements* (ISSN 0971-9091), March 2004.

US Environmental Protection Agency. 1989. *Decision-Maker's Guide to Solid Waste Management, Vol I*. Washington.

US Environmental Protection Agency. 1995. *Decision-Maker's Guide to Solid Waste Management, Vol II*. Washington.

http://ces.iisc.ernet.in/energy/SWMTR/TR85.html

REFERENCES

Freeman, M.H.1988. *Standard Handbook of Hazardous Waste Treatment and Disposal*. McGraw-Hill Book Company, New York.

Phelps, H.O., Heinke, G.W., Jonker, J.F., Ouano, E.A.R. and Vandecasteele, C.1995. *Management of Solid Wastes*. UNESCO, Paris.

Tchobanoglous, G., Theisen, H. and Eliassan, R. 1977. *Solid Wastes—Engineering Principles and Management Issues*. McGraw-Hill Book Company, New York.

http://www.ehs.umaryland.edu/waste/pollutio.htm
http://www.dep.state.pa.us/dep/deputate/airwaste/wm/HW/guides/HWIDGUIDE.PDF
http://www2.kumc.edu/safety/kdhehw/hwg1_10.html#SectionIV

UNIT 10

Integrated Waste Management (IWM)

STRUCTURE

OVERVIEW

From our discussion in Units 1 to 9, it is clear that a single method of waste disposal cannot deal with all waste constituents in an environmentally sustainable way. That is to say, we need a range of management options to achieve the desired results. But, the use of different options (e.g., composting or material recovery) itself depends on functional elements such as the collection and subsequent sorting system employed. In other words, a solid waste management (SWM) system is an integrated system, built on many closely related processes. Instead of focusing on and comparing one option with another (e.g., composting versus landfill), therefore, it is better that we make an attempt to synthesise a waste management system that can deal with the whole waste stream, and then compare their overall performances in environmental and economic terms. This holistic approach is referred to as *integrated waste management* (IWM), and this, in essence, is the focus of this Unit. Integrated waste management (IWM) systems combine waste streams, and waste collection, treatment and disposal methods with the objective of achieving environmental benefits, economic optimisation and social acceptability. Such a waste management system is environmentally, economically and socially sustainable and is likely to be integrated, market oriented, flexible and operated on regional scale.

Against this backdrop, in this Unit, we will examine the overall waste management system and develop ways of assessing overall environmental burdens and economic costs. We will first introduce you to some critical aspects of IWM such as the characteristics, planning and management, benefits and modelling. In order to assess the economic impact, we will then discuss the life cycle assessment (LCA) technique, which examines every stage of the life cycle of solid wastes – from raw material acquisition, through manufacture, distribution, use and possible reuse/recycling to final disposal. (Note that Unit 6 of *Environmental Management* exclusively deals with LCA.) We will then discuss the need for involving the public in waste management activities.

LEARNING OBJECTIVES

After completing this Unit, you should be able to:

- discuss the concept of integrated waste management;
- carry out economic valuation of waste management systems using the LCA technique;
- plan and participate in public education programmes;
- develop an integrated waste management system for your locality involving the public.

10.1 BASICS OF IWM

Waste are by-products of human activity. Physically, wastes contain the same material found in useful products, differ only in their lack of value. The lack of value in many cases can be attributed to the unknown and mixed composition of waste. Separating the materials in waste will generally increase their value, if the separated (i.e., recovered materials) have any use.

- Health and safety have been, and still remain, the major concerns in waste management and, therefore, waste must be managed in such a way as to minimise risks to human health. Today, the society demands a waste management system that is not only safe but also environmentally and economically sustainable. Note that sustainability refers to development that addresses the needs of the present generation without jeopardising the ability of future generations to meet their own needs.

To elaborate, economic cost of a waste management system used to be a major controlling factor in the decision-making process. However, of late, environmental concerns also contribute to the shaping of decisions regarding waste management. Environmental concerns fall under the following two major areas:

(i) **Conservation of resources:** This needs a general reduction in the amount of waste generated i.e., waste minimisation or waste reduction, as well as material and/or energy recovery from the waste. Recovery of resources from waste should slow down the rate of depletion of non-renewable resources and help lower the use of renewable resources to the rate of replenishment.

(ii) **Environmental pollution:** This is the basis for most current environmental concerns about waste management. Levels of emission of pollutants have increased with rise in human population and their activities, but natural processes do not have sufficient turnover to prevent changes in environmental conditions. In extreme cases of overloading, natural processes may break down completely, leading to drastic changes in environmental quality. Along with this, another specific concern emerges at the local level whenever new facilities for waste treatment are proposed. For example, a planned incinerator raises concerns about the increase in emission levels of dioxins, a landfill raises concerns about the risk of landfill gas and groundwater pollution from the leachate generated in landfill sites, etc.

Waste management must, therefore, be (McDougall et al., 2001):

- economically affordable;
- socially acceptable;
- environmentally effective and safe.

To achieve both environmental and economical sustainability, an integrated waste management (IWM) system is necessary. IWM helps in achieving environmental objectives using economically sustainable systems tailored to the specific needs of a community or region.

Before you read any further, let us work through Learning Activity 10.1.

LEARNING ACTIVITY 10.1

Justify the need for an integrated waste management (IWM).

Note:

a) Write your answer in the space given below.

b) Check your answer with the one given at the end of this book.

IWM involves using a combination of techniques and programmes to manage the municipal waste stream. It is based on the fact that waste stream is made up of distinct components that can be managed and disposed off separately. An integrated system is designed to address a specific set of problems, and its operation is based on local resources, economics and environmental impacts. A well-designed IWM plan can improve system economics and reduce environmental impacts, while fostering public support and involvement in solid waste management (EPA, 1989 and 1995). There is no universal step-by-step method for selecting and developing IWM components and systems. That is to say, the success of any IWM depends largely on the dedication and committed expertise of the decision-makers.

Figure 10.1 below shows a hierarchy of waste management that can be used to evaluate the components of IWM vis-à-vis the needs of the community.

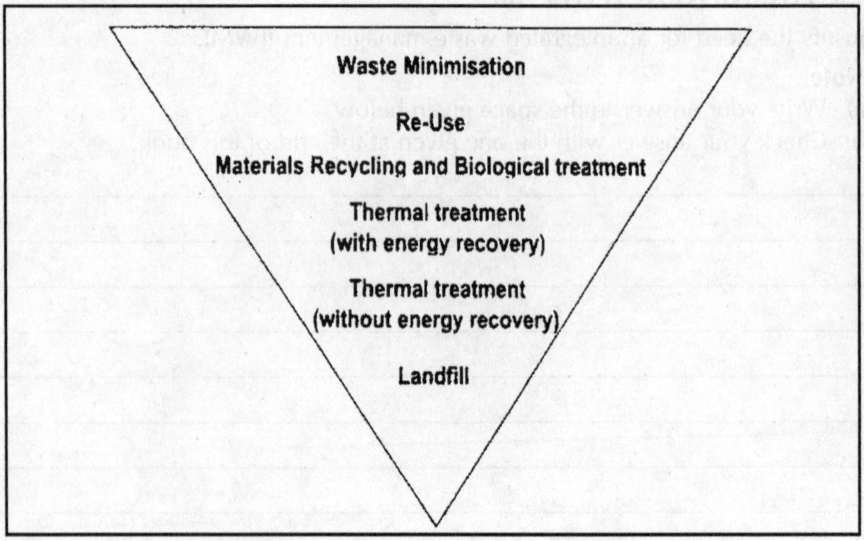

Figure 10.1 Hierarchy of Waste Management

This conceptual tool is useful for goal setting and planning, and all the elements of the hierarchy are interrelated and can be designed to complement each other (McDougall et al., 2001).

Besides the hierarchy of preferred waste management options, a holistic approach illustrated in Fig. 10.2 recognises that all options have a role to play in IWM.

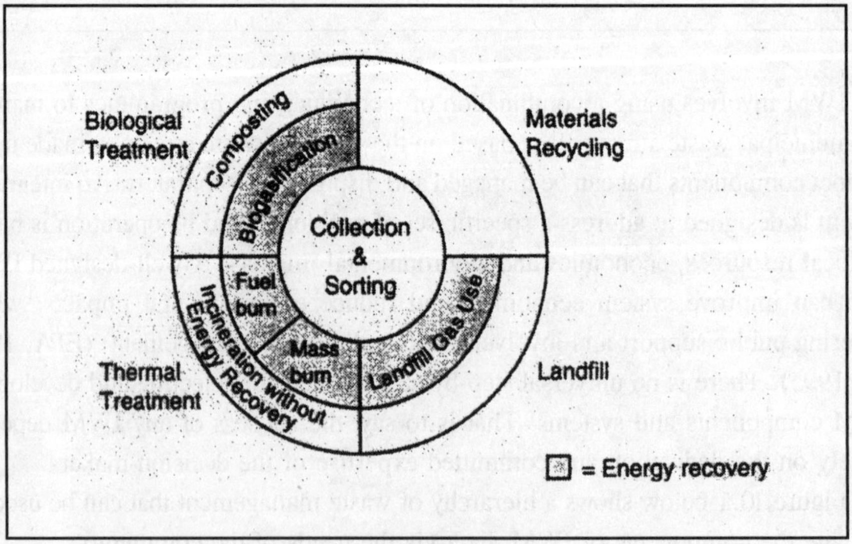

Figure 10.2 Elements of Integrated Waste Management

The above model (Fig. 10.2) illustrates the interrelationships among the parts of the system – the percentage of waste treated by each of the four options will depend on the local conditions. Each option should be assessed using the most recent data available, but the overriding objective is to optimise the whole system, rather than its parts to make it environmentally and economically sustainable and socially acceptable.

Essentially, the overall objective of waste management should be to reduce the amount of waste generated and to manage the waste in a sustainable way by minimising the overall environmental burden associated with the waste management system. However, in one way or another, waste generation is bound to continue as long as there are living organisms, and by extension, waste management is an ongoing process and not a one-time affair. This necessitates the compilation and evaluation of the waste generated and the potential environmental impacts throughout its life cycle. Life cycle assessment (LCA), an environmental management tool, can be used for this purpose. (We will explain LCA later in Section 10.3.)

10.1.1 Characteristics of IWM

The characteristics of IWM include the following:

(i) **An integrated system:** An integrated system combines an optimised waste collection system and effective sorting, followed by one or more of the options, i.e., material recycling, biological treatment, thermal treatment and sanitary landfill. Put differently, an integrated system has control over all types of solid waste materials such as plastic, paper, glass, iron, aluminium, etc. Focusing on a specific material, either because of their ready recyclability (e.g., aluminium) or because of public profile (e.g., plastic), is likely to be less effective in both environmental and economic terms. An integrated system also means control over all sources of solid waste, be they commercial, municipal or domestic. Focusing on a specific source of material (e.g., packaging, domestic or industrial waste) is likely to be less productive than focusing on the nature of the material, regardless of its source.

(ii) **Market oriented:** Any scheme that incorporates material recycling and biological or thermal treatment technologies must recognise that effective recycling of material and production of compost and energy depend on markets for these outputs. These markets are likely to be sensitive to price and consistency in quality and quantity of supply. While working with such schemes, it is important to build markets for outputs, work with secondary material processors and help set material quality standards.

(iii) **Flexibility:** An effective scheme will need the flexibility to design, adapt and operate its system in ways which best meet the current social, economic and environmental conditions. These are likely to change over time and vary by regions. Using a range of waste management options in an integrated system gives the flexibility to channel waste via different treatments, as economic or environmental conditions change. For example, paper can be recycled, composted or incinerated with energy recovery. The options used can be varied according to the economics of paper recycling, compost production and energy generation at the time.

(iv) **Socially acceptable:** For waste management systems to operate effectively, public participation is necessary. Low participation rates in recycling and other schemes can be improved by effective communication. Objection to waste management facility siting can be minimised through appropriate design of facility, and public consultation and education. Where system changes are necessary, effective communication material will inform the public about the benefits of the new system (environmental, economic and social) and thereby increasing the chance of that system being accepted. Public participation in the waste management facilities will also have an effect on the acceptance of the overall waste management system. The integrity of the waste management system is also essential. The public must be confident that any material they source-separate for recycling is sent for recycling and not landfilled or incinerated. Good public support is as essential to a waste management system as good planning and effective management. (We will deal with these aspects in Section 10.4.)

10.1.2 Planning for IWM

The strategic planning for IWM refers to the concept that we must plan for a long term, and the planning process should involve anticipated changes that are likely to occur in the future. It is crucial to build flexibility into all elements of waste management systems. To reiterate, IWM is an ongoing process that has no set beginning or end. The review of new alternatives and evaluation of operations should be performed continually, and planning, development, monitoring and evaluation of options take place simultaneously. Various steps in the IWM planning must follow a conceptual outline discussed below (EPA, 1989 and 1995):

(i) **Organise the framework:** Building local expertise is the primary goal of waste management. Many decision-makers are unfamiliar with the details of various waste management alternatives, even though the issue may not be technical or complex in nature. Investigating and implementing some low technology waste management options such as source reduction programmes, pilot scale recycling, etc., can help them develop their own expertise in areas that may have been previously unfamiliar. Building expertise will better prepare them for implementing larger programmes and will lower the risk of making costly planning mistakes.

(ii) **Understand the institutional and regulatory climate:** It is important to understand and account for local opportunities and constraints when planning a waste management system. The planning should be done according to the political, institutional and economic realities of both local communities and neighbouring political jurisdictions. The knowledge of laws and guidelines affecting municipal waste planning is equally important.

(iii) **Address local waste management issues:** The most fundamental planning factor is to understand the nature of solid waste and its management in the local community. This requires total waste stream assessment, and after having come to grips with all the factors involved, goals and objectives for its management can be decided.

(iv) **Evaluate waste management alternatives:** Evaluating alternatives is the most time consuming activity while developing IWM plans. Dozens of options must be compared and evaluated and feasibility of each option within the local constraints determined.

(v) **Foster public education and involvement:** The involvement of citizens who have been educated about the benefits of proper municipal waste management can be one of the most beneficial aspects of an IWM system. For many municipal waste management options, public participation is one of the main keys of success. Through public education and involvement, one can create opportunities for community members to be a part of the solution to the solid waste dilemma.

(vi) **Understand project financing:** Project financing can be performed in a variety of ways within the community and the selection of financing method is based largely on the degree of risk that the community is willing to take, as the financing method selected can significantly impact the cost incurred by the community..

(vii) **Evaluate new waste management alternatives:** Evaluating new waste management alternatives are necessary to ensure that the local system is as successful as possible. Even the most successful and innovative waste management programmes experiment with new techniques and technologies. Integrated waste management, being an ongoing process, requires continuous attention.

10.1.3 Implementing IWM

As a community moves towards programme implementation, managers must constantly remind themselves to keep the overall programme in perspective. By viewing the project as a whole, no individual element should be paid too much or too little attention. Programme momentum should be sustained at a slow, but steady pace. Issues that can delay or derail a programme must be recognised and appropriately dealt with. Public support must be fostered and confidence in the ability of the community to successfully implement a programme ensured.

To keep a waste management programme in its proper perspective, attention must be paid to the following, popularly referred to as the five "Ps":

(i) **Planning:** Although, it may seem obvious that planning is needed to implement a successful programme, in practice, the need to formulate and follow a well-devised and comprehensive plan is sometimes forgotten. A leaking landfill or other waste management problems may pressure a community to act quickly; hasty actions cause mistakes, which in turn result in delays and wasted resources. While all possible situations cannot be anticipated, many good models, based on successful programmes, do exist, and programme developers are encouraged to use them, when possible, to formulate their own programmes. Planning is especially important because of the potentially large number of factors in the waste management process. Political bodies, waste generators, waste haulers, regulatory agencies, construction contractors, plant operators, energy and material buyers, landfill site owners and citizens must be included for a programme to be successful. Each group has the potential for contributing to or delaying or derailing a project. By formulating and continually reviewing a project plan, programme managers can minimise the chances of a major component of the programme being missed.

(ii) **Price:** Decisions regarding the adoption of alternative strategies for managing waste must continually be based on sound economic analysis that considers the resources of the community and the anticipated environmental impacts and benefits. The community is usually willing to support higher cost waste management options as long as there is confidence that the programme is well run, economically efficient and environmentally sound. Comparing costs and benefits against action is essential for long-term success.

(iii) **Publicity:** Successfully implementing a waste management programme can take a number of years and a commitment of community resources. Support for a programme can erode quickly, unless attention is paid to keeping the programme on the public agenda and maintaining strong and positive public support. A comprehensive plan for informing the public about the programme's progress should be developed and implemented at all stages. Special effort should be made to generate public support before elected bodies vote on programme expenditures. The programme must be seen by the public as something to be proud of, as an example of the progressiveness of the community and its commitment to a clean environment.

(iv) **Political support:** Sustained political support during the long and costly implementation process is vital to the programme's ultimate success. Political support is often crucial to obtain finance and to ensure that the programme gets the resources needed to construct facilities and operate them efficiently. Political leaders should also be kept informed of the programme's progress on a regular basis so that political support for the programme grows as the decision-making body reaches the point of actually committing its public or private resources to implementing the long-term programme. Newly elected political officials must also be informed about the community effort.

(v) **Perseverance:** A community waste management programme must be prepared for the long term. Some projects can take five to ten years to implement and are complex and expensive. A community choosing to implement a programme must be willing to commit the necessary resources to see it through. The ultimate key to success is the will to persevere until the programme is in place.

10.1.4 Benefits of IWM for Developing Economies

The management of municipal solid waste (MSW) is an integral, but much neglected, aspect of environmental management in many developing countries. The waste management systems, where already in place, in a majority of the developing economies are often characterised by inadequate collection services, little or no treatment and uncontrolled dumping. Thus, a lack of an effective mechanism for waste management and infrastructure coupled with severely limited resources warrant a simple, but robust, IWM system in these countries. The establishment of IWM system will require the following (McDougall et al., 2001):

- Data on waste composition. Collection of reliable data, considered the foundation of IWM, is required for effective planning of collection, transport and treatment of MSW.
- Progress from uncontrolled dumping to the use of simple sanitary landfills.
- Separation of organic waste, which can be composted, from MSW.
- Institutionalised involvement of personnel in the separation and collection of recyclable materials.

Given the available resources and MSW composition, the waste hierarchy for waste management (See Fig. 10.1) is evidently too rigid to be relevant to developing countries. In other words, the characteristic flexibility of IWM offers a more realistic opportunity to improve waste management by addressing local conditions.

Although technical and financial resources are limited, developing economies still have the potential to significantly improve waste management. For example, moving from open dumping to simple sanitary landfill in conjunction with separation and composting of organic waste is likely to result in significant benefits. This prevents, or at the very least minimises, the pollution of surface and groundwater by leachate, migration of combustible gases (methane), odour and breeding of disease carriers. The institutionalised involvement of scavengers in the collection, sorting and recycling of materials helps improve their living conditions, bring to them dignity, reduce the health risks they are exposed to, supplement their income and increase recycling rates. A careful analysis of market conditions for recyclable materials and compost could be conducted to prevent imbalances that could affect their final prices. In essence, an IWM can provide an environmentally effective, economically affordable and socially acceptable waste management system.

10.1.5 Waste Management Modelling

Optimising the waste system to reduce environmental burdens or economic costs requires the burdens and costs to be predicted. Put differently, modelling of waste management systems is necessary. Several models can be produced for predicting environmental burdens. This is a new branch of applied science, which has resulted in the development of a new environmental management tool. The modelling technique used very often is life cycle inventory (LCI) of solid waste, which we will explain in Section 10.2.

Modelling in general can be divided into the following two areas:

(i) model structure, which will determine how the model will work;
(ii) data acquisition for verification and validation of the model (McDougall et al., 2001).

Even the best models are useless without accurate, relevant and accessible data. Through the efforts of a range of agencies/bodies, many different pilot schemes, both on small and large scales, have been set up and data are becoming available, because of which it is now possible to model IWM schemes, based on actual data.

A word of caution is necessary here. Modelling may sound a purely academic exercise, but in reality, it has several practical uses, some of which are listed below:

- The process of building a model focuses attention on missing data. Often the real costs, in either environmental or economic terms, for part of the waste chain are not widely known. Once identified, the missing data can either be sought out, or, if not existing, analysis can be carried out to gather the relevant data.
- Once completed, a model will define the current status of waste management, both by describing the system and by calculating the overall economic cost and environmental burden.
- A model can also be used to predict environmental burdens and likely economic costs in the future. Such a forecast may not be 100% accurate, but will give a workable estimate to work out a future strategy.
- Modelling allows other calculations to be made, which can then be used to define the sensitive points in the system. This will show the changes that will have the greatest effects in reducing costs or environmental burdens.
- Modelling the waste system will allow prediction of the likely amounts of reclaimed material available, which will, in turn, allow informed investments in the necessary equipment.

✍ LEARNING ACTIVITY 10.2

List the steps involved in implementing IWM.

Note:

a) Write your answer in the space given below.
b) Check your answer with the one given at the end of this book.

We will now discuss the role of life cycle assessment in solid waste management. You may also refer to Unit 6 of the **Environmental Management** Course for a detailed discussion of LCA techniques.

10.2 LIFE CYCLE ASSESSMENT (LCA)

Life cycle assessment (LCA) is a compilation and evaluation of the inputs, outputs and the potential environmental impacts of a product system throughout its life cycle. It is an environmental management tool increasingly used to understand how a product or service is provided from 'cradle to grave' (McDougall et al., 2001). That is to say, the technique examines every stage of the life cycle – from raw material acquisition through manufacture, distribution, use, possible reuse/ recycling to final disposal. In addition, every operation or unit process within a stage is included and for every operation within a stage, the inputs (i.e., raw materials, resources and energy) and outputs (i.e., emission to air, water and solid waste) are calculated. These inputs and outputs are then aggregated over the life cycle. The environmental issues associated with these inputs and outputs are then evaluated in the life cycle impact assessment (LCIA). Conducting LCAs for alternative products or services thus allow for improved understanding and comparisons.

10.2.1 Benefits of LCA

Life cycle assessment is an inclusive tool. The life cycle inventory phase is essentially an accounting process or mass balance for a system. All necessary inputs and emissions in many stages and operations of the life cycle are considered to be within the system boundaries. This includes not only direct inputs and emission for production, distribution, use and disposal, but also the indirect inputs and emissions. It is essential that all the processes are included within the boundaries to conduct a fair and transparent analysis. LCA offers the prospect of mapping the energy and material flows as well as the resources, solid wastes and emissions of the total system, i.e., it provides a system map that sets the stage for a holistic approach. Comparing such system maps for different options, whether for different products or waste management systems, allows the identification of areas, where environmental improvements can be made.

Looked at from a different standpoint, the concern about the environment is sometimes expressed in terms of individual issues, such as acidification. The power of LCA is that it expands the debate on environmental concerns beyond a single issue and attempts to address a broad range of environmental issues. The system map also allows other environmental information and assessment tools to be incorporated in conjunction with LCA.

10.2.2 LCA Phases/Stages

LCA consists of the following four phases or stages:

(i) **Goal definition and scope:** At this phase, the functional unit for comparison is defined as well as the study purpose, system boundaries, life cycle stages, unit processes and scope of assessment.
(ii) **Life cycle inventory analysis (LCI):** This involves the process of accounting for all the inputs and outputs of the product system over the life cycle, resulting in a list of raw material and energy inputs, and individual emission to air, water and solid waste.
(iii) **Life cycle impact assessment (LCIA):** This phase associates the inputs and outputs with particular environmental issues, e.g., ozone depletion, and converts the inventory of material, energy and emissions into representative indicators, e.g., an aggregate loading of ozone depleting chemicals.
(iv) **Life cycle interpretation:** This phase evaluates the significance of the inputs, outputs and indicators of the system life cycle. (Note that this stage is the least accepted or defined.)

Figure 10.3 illustrates the four-phase structure of LCA.

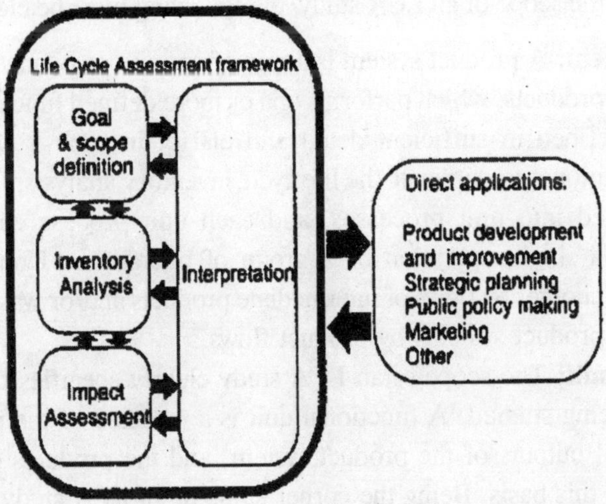

Figure 10.3 LCA Phases

Source: McDougall et al., 2001

Referring to Fig. 10.3, in what follows we will discuss these phases/stages in detail.

Goal and scope definition

The goal of an LCA study must unambiguously state the intended application, objectives and intended audience. It should state the reason for performing a specific study and define the options that are to be compared and the intended use of the result. The intended use will influence the type of study and the type of data required. This stage also involves identifying the system boundaries (i.e., technical, geographical and time) and the procedures to handle the data. Rules and assumptions must be documented, especially with respect to allocation rule for co-products, open loop recycling and aggregation. All the decisions that need to be taken, while carrying out an LCA should be made in light of the purpose of the study. A clearly defined goal will partly address the need for transparency and will help ensure whether or not the result is fit for use.

Transparency is essential throughout the life cycle procedure. This allows data tracking and calculation verification to be carried out, if necessary. A complete and transparent record of a study is compiled in a final study report. The scope of study should be sufficiently well defined to ensure that the breadth, depth and details of the study are compatible and sufficient to address the stated goal.

In defining the scope of an LCA study, the following must be clearly described:

- **Product system:** A product system is a mix of operations connected by flows of intermediate products, which performs one or more defined function. The system should be defined in sufficient detail and clarity in such a way as to allow another practitioner to duplicate the life cycle inventory analysis. Product systems are subdivided into unit processes, and each unit process encompasses the activities of a single operation or a group of operations. Unit processes are linked to one another by flows of intermediate products and/or waste for treatment and to other product systems by product flows.
- **Functional unit:** The scope of an LCA study clearly specifies the functions of the system being studied. A functional unit is a measure of the performance of the functional outputs of the product system, and the products or services are compared on this basis. Being the corner stone of an LCA study, the functional unit provides the reference point to which both inputs and outputs are related and allow a clear comparison of LCA results. The stages and unit processes connected to the functional unit are known as the product system, which we described above.

- **System boundaries:** The system boundary defines the unit processes that will be included in the system for modelling purposes. Ideally, the product system should be modelled in such a manner that inputs and outputs at its boundary are *elementary flow*, i.e., the material/energy entering the system under study has been drawn from the environment without previous human transformation, or the material/energy entering the system under study is discarded into the environment without subsequent human transformation. The system boundaries, i.e., the stages, operations, inputs and outputs included and omitted within the assessment, need to be defined. In any LCA, defining the functional unit and the boundaries of the system being assessed are both important steps. Reporting the functional unit and the boundaries used needs to be clear and detailed with any omissions stated and justified. This is important in giving the actual results.

Life Cycle Inventory (LCI) Analysis

A life cycle inventory analysis is concerned with the data collection and calculation procedures necessary to complete the inventory. This stage accounts for constructing a material and energy balance for each step in the life cycle. The analysis of all inputs and outputs for each stage in the life cycle can then be combined to give the overall life cycle inventory of the product or service. The procedure, as shown in Fig. 10.4, entails describing the life cycle as a series of steps and then calculating the inputs and outputs for each of these steps.

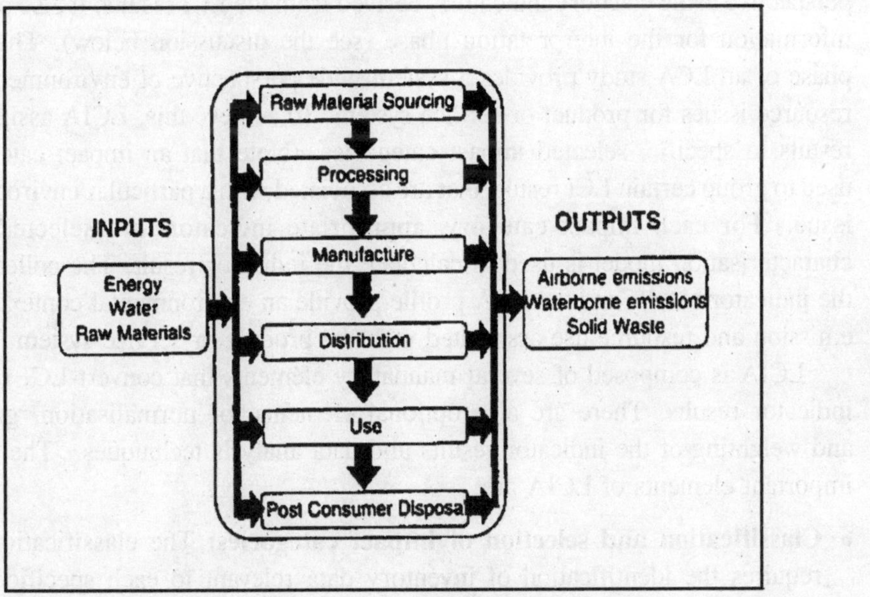

Figure 10.4 The Stages of a Product's Life Cycle

An LCI provides enormous knowledge and insights into the operation of a given system, which can provide the basis for applying and integrating other environmental information and assessment tools into a system for comparison. For LCI, the main categories of data required are:

- specific data for production, distribution and waste management;
- generic data for energy production, raw material transportation and extraction.

Data quality requirements should address time related, geographical and technological issues, the precision, completeness and representativeness of the data and the consistency and reproducibility of the methods used throughout the LCA. LCI should also undergo sensitivity and uncertainty analysis. Sensitivity analysis can help identify whether any of the assumptions made have a significant influence on the outcome of the LCI, and if so, which assumption has the greatest influence. Uncertainty analysis is essential as it identifies those parameters for which the margins of uncertainty have a large influence on the result, and subsequently, the margins of uncertainty for these parameters can be improved. The data and results should not be used without understanding their quality and limitations.

Life Cycle Impact Assessment (LCIA)

The LCIA phase aims to examine the product system from an environmental perspective using category indicators, derived from the LCI results. It also provides information for the interpretation phase (see the discussion below). The LCIA phase of an LCA study provides a system wide perspective of environmental and resource issues for product or service system. To achieve this, LCIA assigns LCI results to specific, selected impact categories. (Note that an impact category is used to group certain LCI results that are associated with a particular environmental issue.) For each impact category, appropriate indicators are selected and a characterisation model is used to calculate the indicator result. The collection of the indicator results and the LCA profile provide an environmental context for the emission and resource use associated with the product or service system.

LCIA is composed of several mandatory elements that convert LCI result to indicator results. There are also optional elements for normalisation, grouping and weighting of the indicator results and data analysis techniques. The various important elements of LCIA are:

- **Classification and selection of impact categories:** The classification stage requires the identification of inventory data relevant to each specific impact category and the assignment of the appropriate LCI results to each category.

Data may belong to more than one category, e.g., NO_x has a global warming and an acidifying effect. The impact categories are selected based on the goals and scope of the LCA study.

- **Characterisation:** The aim of characterisation is to provide a basis for the aggregation of the inventory results into an indicator for each category. Each impact category requires a specific model to convert the inventory results into the indicator. For example, in case of global warming, the most common indicator used is Global Warming Potential (GWP) in CO_2 equivalents. There are two steps in this calculation – each greenhouse gas is first converted into carbon dioxide equivalents based on particular characterisation factors and the individual carbon dioxide equivalents are then added into a total indicator.
- **Normalisation:** This involves relating the characterised data to a broader data set or situation, as in, for example, relating the SO_x emission to a country's total SO_x emission. Normalisation can provide insights but should be treated with caution, as results can differ significantly, if different data sets are used. Normalisation is often omitted from LCA studies.
- **Weighting:** This is the process of converting indicator results of different impact categories into scores by using numerical factors based on values. Weighting may include aggregation of the weighted results into an overall score. This is the most subjective case of an LCA, which is based on value judgement and is not scientific. Different individuals or organisations may have different preferences or values. Different parties, therefore, are likely to reach different weighting results based on the same indicator results.

Life cycle interpretation

Life cycle interpretation is a systematic technique to identify, qualify, check and evaluate information from the results of the life cycle inventory (LCI) analysis and/or LCIA of a product system, and present them in order to meet the requirements of the application as described in the goal and scope of study. The interpretation stage of the LCA process is closely linked to the iterative nature of the process of scope definition, inventory analysis and impact assessment. Interpretation involves a review of all the stages in the LCA process.

There are a few sub-phases at this stage of an LCA, and these are:

- **Identification of significant issues:** Based on LCI and LCIA phases of LCA, the objective of this sub-phase is to structure the results from LCI or LCIA phases in such a way as to determine the significant issues, including any implications of the particular method used and assumptions made, allocation rule, cut-off decisions, choice of indicators and characterisation.

- **Evaluation:** The objectives of the evaluation sub-phase are to establish and enhance the confidence and reliability of the result of the study. The result of evaluation should be presented in such a way as to allow the reader a clear and understandable view of the outcome of the study. To achieve this, a completeness check (ensuring that all relevant information for interpretation is available and complete), a sensitivity check (assessing the reliability of the results by assessing the uncertainty of the significant issues affecting the conclusion) and a consistency check (determining whether the assumptions, methods and data are consistent with the goal and the scope) must be carried out.
- **Conclusions, recommendations and reporting:** Drawing conclusions from a study should be done interactively with the other elements in the life cycle interpretation phase. Preliminary conclusions must be checked to ensure that they are consistent with the goal and scope of the study. Recommendations should be based on the conclusions of the study and the final report shall present a complete, unbiased and transparent account of the whole study.

Note that an LCA will allow the trade-offs associated with each option to be assessed, which is economically affordable, socially acceptable and environmentally effective.

While LCA, used with other environmental information and assessment tools, helps improve the environmental performance of a product or service, public involvement and education is necessary for a sustainable solid waste management system. We will study this next in Section 10.3. But, first, let us work out Learning Activity 10.3.

✍ LEARNING ACTIVITY 10.3

List the stages of LCA.

Note:

a) Write your answer in the space given below.

b) Check your answer with the one given at the end of this book.

10.3 PUBLIC EDUCATION AND INVOLVEMENT

Developing integrated solutions for waste management problems requires public involvement. To operate a waste management programme economically and efficiently, significant cooperation is expected from waste generators, regardless of the strategies chosen – buying products in bulk, separating recyclables from non-recyclables, dropping off garden trimmings at a compost site, removing batteries from materials sent to a waste-to-energy facility, or using designated containers for collecting materials. To sustain a long-term programme support, the public needs to know clearly the purpose and necessity of desired behaviour.

Involving people in the waste management requires a significant educational effort by the community. Ineffective or half-hearted education programmes may confuse the public, reduce public confidence or elicit hostility towards the programme. Successful education programmes must be consistent and ongoing. Public education stimulates interest in how waste management decisions are made, and when citizens become interested in their community's waste management programmes, they frequently demand to be involved in the decision-making process. Communities should anticipate such interest and develop procedures for involving the public. When the public is involved in programme design, it helps ensure the smooth running of the programmes (EPA, 1989 and 1995).

We now discuss some aspects of public education and involvement in Subsections 10.3.1 and 10.3.2.

10.3.1 Planning a Public Education Programme

Successful public education plans are the result of careful planning. By developing a realistic education plan, you can assess the situation and know where best to direct their efforts and resources. This will benefit from taking advantage of all opportunities to work with the community. The process of developing an education and involvement plan provides an opportunity to involve the community in the planning process at an early point. The education plan must begin by introducing people to waste management needs and concepts, explaining clearly how to participate, and then effectively encouraging them to adopt the desired waste management behaviour.

Once the public participates in the programme, incentives and reinforcements can be used to maintain and increase participation rates. Developing an effective education programme requires planning and research. Programme developers must use different strategies for different groups, such as homeowners, apartment dwellers, business people and school children. Grounded on a sound information base, an effective education programme moves people through a series of stages (EPA, 1989 and 1995). Let us describe each of these stages now.

Awareness

At the awareness stage, people encounter a new idea or a new way of doing things. At this stage, they do not possess enough information to decide whether a change in behaviour is a good idea or whether they should be concerned. The goal of the awareness stage is to let people know that a different way of handling waste may be preferable to the historical way and that good reasons for considering a change in their waste management practices do exist. A variety of methods can increase awareness, in which low-cost methods include news articles and public service announcements, or shows on radio and television and high cost efforts include television commercials or billboards. Over the long term, education in schools is the best way of raising awareness; the school curriculum must include the concepts of source reduction, recycling composting and other waste management techniques. Besides educating the next generation of citizens, school programmes indirectly help make parents aware of waste issues because children frequently take home information they have learned and discuss it with their parents.

Interest

In the second stage, individuals who are now aware of waste management issues seek additional information. They may seek one-to-one exchanges with waste management professionals, educators, and so on or they may seek information about how they are involved in implementing a waste management initiative or an effective public policy. Making changes in required local waste management practices, such as mandatory recycling or garden trimmings disposal bans, will clearly stimulate interest (sometimes even in the form of political opposition!). At this stage, programme developers may need a variety of methods to explain the programme. Making public speeches, offering tours of waste management facilities, creating exhibits for fairs and preparing written material such as newsletters can help stimulate public interest in the programme.

Evaluation

At the evaluation stage, individuals decide whether to go along with the programme. Even if the law requires specific behaviour, achieving voluntary compliance is easier administratively and politically than strong enforcement. An easily understandable and convenient programme will have the best chance of success. Research has shown that for even well promoted programmes, initial participation is about 50%. Initial high participation rates, therefore, should not be expected. Even for mandatory programmes, convenience is a major factor in determining participation. For example, the convenience of curbside pickup normally makes participation in waste management programmes higher than for drop-off programmes. As a result, some communities only provide drop-off service for garden trimmings, so that it becomes more convenient not to collect grass clippings or home compost. A combined curbside and drop-off programme may be the most convenient. To make this happen, education should stress the role of individuals in the programme, their contribution to its success and the most convenient level of participation.

Trial

By the fourth stage, individuals would have decided to participate in the new activity. This is a crucial step for every programme. For example, if individuals try garden composting or a volume-based system and encounter difficulty, they may choose not to adopt the desired conduct, and the programme could lose political and public support. By this stage in the educational programme, therefore, everyone should have the information, describing exactly what they are expected to do. The community programme must then provide the promised service in a highly reliable fashion. At the trial stage of a volunteer programme, a pilot project can also help stimulate participation. Programme organisers should assure citizens that the pilot project's goal is to evaluate various strategies, respond to public feedback and make any changes required to improve programme efficiency and reliability. Citizens may be more willing to try a project, if they know whether the project is short-term and that any concerns they may have will be taken into account in developing a long-term effort. During the trial stage, public hearings may be helpful by giving citizens an opportunity to voice their opinions about the project.

Adoption

If the education programme has been well planned and implemented, public support and participation should grow. Educational efforts at this fifth stage focus on providing citizens with positive feedback concerning programme effectiveness. A newsletter or other regular informational mailing can help inform citizens about the programme's progress and any programme changes. Community meetings can serve to reward and reinforce good behaviour and answer questions. Local officials should be informed of programme participation rates to generate political support for programme budgets and personnel needs.

Maintenance

At the sixth stage, the programme progresses smoothly. The use of a variety of intrinsic and extrinsic incentives will maintain and increase participation. Intrinsic incentives are largely information in nature, designed to induce citizens to perform the desired conduct and to provide a personal sense of well-being and satisfaction. Some studies, for example, have shown that the ideals of frugality, resource conservation and environmental protection over the long run were strong intrinsic motivators for those participating in recycling and reuse programmes. Extrinsic incentives are tangible and direct rewards for performing the desired conduct, such as reduced fees or monetary payments. For example, the smaller the waste volume generated, the lesser the generator (of wastes) must pay for waste management. A maintenance programme may employ both the types of incentives, while continuing with basic education.

10.3.2 Planning Public Involvement

Participation of local residents should begin at the very early stage of the programme development. For example, it should begin even at the stage of making decisions regarding the overall waste management strategy that best meets the community's economic and environmental needs. The strategy should consider source reduction and other options in addition to the facility being proposed. The public must also accept responsibility for its role in implementing sound and cost effective waste management solutions.

Developing a written plan for seeking public involvement is important. Written procedures help ensure the inclusion of all important interests and legal requirements. The plan will show involved citizens and groups at which points in the process they can express opinions and how to be most effective in communicating their views. In fact, a written, publicly available plan lends credibility to the programme. The "issue evolution-educational intervention" (IEEI) model provides public involvement throughout the decision-making process (EPA, 1989 and 1995). This model comprises an eight-stage process for developing and implementing public policy. The IEEI process ensures that the public will have a meaningful voice in deciding how best to manage solid waste. The process is not simple, and requires a commitment from the community for time and resources. Each of these stages is described below:

(i) **Concern:** In the first stage, an event puts waste management on the public agenda. The public begins to ask questions. At this stage, a procedure for providing accurate and reliable information to the public is important. Eliminating misconceptions and establishing a firm educational base for public discussion is the key. Educational institutions, offices, governmental associations and regulatory agencies can provide information. Education programmes should target local officials, as well as the public. Showing concern and a willingness to take proper action is most important. A focus group can help define important public issues, and a community service organisation can provide a forum for discussion.

(ii) **Involvement:** As discussion of the issue begins, regulatory officials, persons from neighbouring communities, local waste management experts, environmental and business groups and others should be encouraged to participate. Bringing representatives of interest groups together and providing a forum for communication is a valuable activity. Cultural diversity is another consideration when seeking input from the broadest possible spectrum of the community.

(iii) **Issue resolution:** Interest groups should make clear their points of agreement and disagreement. The various groups should then attempt to understand and resolve points of conflict. Determining what people can agree on is also important. All parties need to understand the motivation and circumstances of the other community interests in the process.

(iv) **Alternatives:** The participants should develop a list of available alternatives and each alternative should have a list of potential sites for facilities. At this stage, participants should use the same criteria to analyse comparative economics, environmental impacts and other aspects of each alternative. Each interest group should scrutinise carefully the analyses prepared by others. Results of analyses of various alternatives should be communicated to local officials and input sought from the public and others.

(v) **Consequences:** At this stage, involved parties should determine and compare the economic and environmental effects of each alternative. They should also evaluate consequences in light of community resources and goals, and the public must understand the results of choosing one alternative over another. All involved interest groups should acknowledge the benefits and costs associated with each alternative.

(vi) **Choice:** The deciding body must decide, at this stage, an alternative or a combination of alternatives chosen for implementation. In addition to publicising the chosen alternative or alternatives, decision-makers should clearly communicate the reasons behind each choice by explaining the necessary trade-offs, the efforts made to consider the interests of each affected group and the anticipated impact of the chosen alternative or alternatives on the community. This will help develop a broad community consensus, enabling the community to better withstand legal and political challenges. (We should not, however, expect a 100% support from all the interest groups involved to the chosen alternative or alternatives!)

(vii) **Implementation:** At this stage, the decision-makers should describe the steps necessary to implement the chosen strategy. They should also try to mitigate potential adverse impacts, which the chosen alternative or alternatives may have on relevant interest groups.

(viii) **Evaluation:** The community should continually evaluate the model and solicit input from affected groups. The impact of decisions should be communicated routinely to the public and to the local officials. Ongoing evaluation helps provide an information base for making future waste management decisions, and the existing programmes will continually improve, if they respond to changing conditions and public input.

✍ LEARNING ACTIVITY 10.4

List the stages to be considered for public participation in the management of solid wastes.

Note:

a) Write your answer in the space given below.

b) Check your answer with the one given at the end of this book.

SUMMARY

In this Unit, we discussed the concept of integrated waste management (IWM) system and said that it aims at reducing the amount of waste generated and managing the waste in a sustainable way by minimising the overall environmental burden. We also pointed out that an integrated system is designed to address a specific set of local solid management problems, and its operation is based on local resources, economics and environmental impacts. We then discussed the characteristics, planning, implementation and various benefits of IWM. Subsequently, we explained the stages involved in, and the relevance of, life cycle assessment (LCA) technique as an environmental management tool. We closed the Unit by stressing the need for public participation to sustain solid waste management activities.

SUGGESTED READINGS

Areivala, S.J. 1971. *Solid Wastes Disposal in India*. Central Public Health Engineering Institute, Nagpur.

Arne Vesilind, William Worrel and Reinhart Debra. 2002. *Solid Waste Engineering*. Thomson Brooks/Cole, Singapore.

Ramachandra, T.V. and Saira Varghese, K. 2003. Exploring possibilities of achieving sustainability in solid waste management. *Indian J. Environmental Health*, **45 (4):** 255-264.

Tchobanoglous, G., Theisen, H. and Samuel, A. Vigil. 1993. *Integrated Solid Waste Management*. McGraw-Hill, Inc., New Delhi.

Environmental Management in Developing Countries, 1995. *Waste Management Vol II*. Institute for Scientific Co-operation, Tubingen.

Flinotoff, R. 1984. *Management of Solid Wastes in Developing Countries*. WHO Regional Publications, Southeast Asia Series No 1.

http://ces.iisc.ernet.in/energy/SWMTR/TR85.html

REFERENCES

McDougall, F.R., White, P.R., Franke, M. and Hindle, P. (2001). *Integrated Solid Waste Management: A Life Cycle Inventory,* Blackwell Science, UK.

US Environmental Protection Agency (1989). *Decision-Maker's Guide to Solid Waste Management, Vol I*. Washington.

US Environmental Protection Agency (1995). *Decision-Maker's Guide to Solid Waste Management, Vol II*. Washington.

Model Answers to Learning Activities

UNIT 1: INTRODUCING MUNICIPAL SOLID WASTE MANAGEMENT

Learning Activity 1.1

I reside in ward no. 89 of Bangalore (Karnataka, India), which is a residential area and has a population of 29,440. It has two colleges, three schools, one hospital and few commercial establishments. Based on source and type of wastes generated in my ward, they can be classified as follows:

(i) Based on Source

Residential: Consisting of single and multi-storied buildings (food leftovers, vegetable peels, plastics, clothes, etc.).

Commercial: Consisting of provision stores, hospital wastes and auto repair shops (food leftovers, glasses, metals, etc.).

Institutional: Consisting of schools and colleges (paper, plastics, glasses, etc.).

Municipal: Consisting of wastes from demolition and construction activities (dust, building debris, etc.).

(ii) Based on Type

- Garbage (food wastes, vegetable peels, etc.).
- Ashes and residue.
- Combustible and noncombustible wastes (plastics, textiles, glass, etc.).
- Bulky wastes (tyres, trees, branches, crates, etc.).
- Street wastes (dirt, leaves, etc.).
- Dead animals (due to natural or accidental death, e.g., dogs, cats, cattle, etc.).
- Construction and demolition wastes (rubber, concrete, plaster, bricks, etc.).

Learning Activity 1.2

Currently, the functional elements in my locality are waste generation, storage, collection and disposal.

Improvements can be made on the existing functional elements by the addition of transport and processing. Since our locality is mostly residential, the quantity of biodegradable waste generated is high. It is, therefore, important that wastes are collected at least three times a week. In order to combat the problems of narrow access ways, a transfer station can be set up, which could be a small area along the street, where collectors transfer wastes from smaller trucks or pushcarts to bigger vehicles. Segregation of wastes can be done at the transfer station, and depending on the waste, they can be disposed off in landfills.

Learning Activity 1.3

Environmentally sound management is an integrated approach for controlling and preserving the resources, both in quantity and quality. Its role is to ensure sustainable development of the ecosystem and human environment by minimising the effects of human activities.

Environmental sound technologies (EST) are cost effective and energy efficient and they do not pollute the ecosystem's vital components such as air, land or water. EST can be categorised broadly as hard and soft. Hard ESTs include equipment, machines and other infrastructure with their material accessories, to handle waste products and monitor and measure environmental quality in the air, water and soil. Soft ESTs support and complement hard technologies and include nature-based technologies and management tools.

Learning Activity 1.4

Waste composition in our locality consists of wet waste and dry waste. Wet waste constitutes 60% of the solid waste generated, e.g., food wastes, vegetable peels, etc. Dry waste constitutes the remaining 40% of the solid wastes, i.e., paper 12%, plastics 14%, glass 4%, rubber 1%, battery and expired medicines 2%, iron 1%, dust 5% and cardboard boxes 1%.

There has been an increase in the number of residential dwellings in our locality. Due to the increase in population, the quantity of wet wastes (e.g., food leftovers, vegetable peels, etc.) and dry wastes (e.g., plastics, paper, etc.) have also increased, but the percentage of generation remains constant. The waste generated per person is 0.35 kg per capita per day.

UNIT 2: WASTE GENERATION ASPECTS

Learning Activity 2.1

The cornerstone of successful planning for a waste management programme is reliable information about the quantity, composition and sources of wastes. Waste stream assessment (WSA) helps in analysing the short- and long-term problems within the local waste management system. WSA is required to identify components that require improvement for implementing waste management programmes. Field investigations for WSA are carried out in the absence of reliable basic data in three ways: (i) sorting of wastes, (ii) vehicle weighing at disposal sites and (iii) visiting institutional and industrial sites to identify wastes being generated and disposal methods.

Waste stream assessment at present is not practised in our locality, and the waste is usually dumped in storage containers and taken to landfill sites.

Learning Activity 2.2

Some of the important factors involved in the variation of composition of wastes are geographic location, seasons, collection frequency, population diversity, extent of salvaging and recycling, public attitude and legislation.

Geographic location is related primarily to the different climates that can influence both the amount of a particular type of solid waste generated and the collection operation. A majority of the developing countries are in the tropical region and thus their composition of wastes differs from that of developed countries. The frequency of waste collection depends on the quantity of waste generated. It is seen that the amount of wastes generated is more in low-income area compared to high-income area. Similarly, the composition differs in terms of paper and other recyclables, which are more in high-income area compared to low-income area. However, public attitude towards wastes will bring about a significant reduction in the quantities of solid wastes, if and when people are willing to change, on their own volition, their habits and lifestyles to conserve the natural resources. The last but not the least is the existence of local and state regulations concerning the use and disposal of specific materials.

Learning Activity 2.3

Component	Percent by mass	Moisture content (%)	Dry mass, kg
Food waste	15	70	4.5
Paper	45	6	42.3
Cardboard	10	5	9.5
Plastics	10	2	9.8
Garden trimmings	10	60	4.0
Wood	5	20	4.0
Tin cans	5	3	4.9
Total	100		79.0

Using the formula:

$$\text{Moisture content} = \frac{\text{Wet weight} - \text{Dry weight}}{\text{Wet weight}} \times 100$$

$$\text{Therefore moisture content} = \frac{(100 - 79)}{100} \times 100 = 21\%$$

Learning Activity 2.4

The information on the chemical characteristics is important in evaluating alternative processing and recovery options. If solid waste is to be used as fuel or used for any other purpose, we must know its chemical characteristics including lipids, carbohydrates, proteins, natural fibres, synthetic organic materials (plastics) and noncombustible materials.

The two types of wastes commonly present in our locality are combustible wastes and garbage. Combustible waste such as plastics consists of natural fibres, and garbage such as food wastes have carbohydrates, lipids and proteins.

Learning Activity 2.5

Some of the adverse health and environmental effects, due to the improper handling of wastes are:

(i) **Health effects**: Wastes dumped indiscriminately provide the food and environment for breeding of various vectors, e.g., flies (salmonellosis, dysentry, etc.), mosquitoes and roaches (malaria, dengue fever, typhoid, cholera, amoebiasis, etc.) and animals, e.g., rodents and pigs (trichinosis, cysticerosis, etc.).

(ii) **Environmental effects**: Inadequate and improper waste management has serious environmental effects. These include air, water, land, visual, noise and odour pollution, and explosion hazards.

I reside in ward no. 89 of Bangalore, Karnataka, India where sufficient precaution is not practised, while handling municipal solid wastes. Based on a general observation, the four effects are the following:

- Due to open dumping, mosquitoes thrive in our locality, which may cause diseases like malaria or dengue fever.
- Rodents, notably rats proliferate in uncontrolled deposits of solid waste, which provide them with a convenient source of food and shelter.
- There is a risk of injury during handling of wastes, as workers are not provided with safety materials, e.g., gloves.
- The aesthetic sensibility (i.e., visual pollution) of concerned residents is offended by the unsightliness of piles of wastes.

UNIT 3: WASTE COLLECTION, STORAGE AND TRANSPORT

Learning Activity 3.1

Bangalore Mahanagara Palike (BMP) undertakes the waste collection system in our locality.

Solid waste collection starts at the point of waste generation. Wastes are stored in bottomless concrete containers placed 100 metres apart for which covers are not provided. Wastes are collected in trucks of 4 to 5 tonnes capacity. Collection frequency is based on the requirement of the locality. Since our locality is mostly residential, collection of waste is done three times a week. The collection crew consists of a driver and two helpers. The collection route for our locality is entirely left to the driver's judgement.

Collection is often the most costly component of the solid waste management system and a proper collection system design and management can reduce the cost significantly. In terms of cost, the collection system in developing countries accounts for 70–80% of the total budget for solid waste management, the remaining 20–30% going for overheads. While making decisions for alternative collection systems, the services must be evaluated considering such factors as efficiency, effectiveness, equity, reliability, safety and environmental impacts.

Learning Activity 3.2

The types of containers used in our locality are uncovered bottomless concrete rings having 0.9 metres diameter and 0.6 cubic metre storage capacity. They are placed 100–200 metres apart. The collection vehicle used is a truck of 4–5 tonnes capacity and is enclosed by an iron mesh at the top to prevent the spillage of wastes.

Learning Activity 3.3

The heuristic (trial and error) route development process is a relatively manual approach that applies specific routing patterns to block configurations. Routes are to be traced on the tracing paper by following certain rules. Routes should not be fragmented or overlapping. Each route should be compact, consisting of street segments clustered in the same geographical area. In streets with heavy traffic, wastes should not be collected during peak hours. Higher elevations should be at the start of the route. Waste on a steep hill should be collected, when practical, on both sides of the street while vehicle is moving downhill. This facilitates safe, easy and fast collection. In case of one-way streets, it is best to start near the upper end of the street, working down it through the looping process. Services on dead end streets can be considered as services on the street segment. To keep right turns at a minimum, collect the dead-end streets when they are to the left of the truck. They must be collected by walking down, backing down or making a U-turn.

Learning Activity 3.4

Transfer station is a centralised facility where waste is unloaded from smaller collection vehicles and reloaded into large vehicles for transport to a disposal or processing site. To determine whether a transfer system is viable for a particular community, the decision-makers should compare the costs and savings associated with the construction and operation of a transfer facility. The use of transfer station is a sound practice when there is a need for vehicles servicing a collection route to travel a shorter distance, unload and return quickly to their primary task of collecting the waste. There are three types of stations depending on the capacity, viz., small capacity (less than 100 tonnes/day), medium capacity (100 to 500 tonnes) and large capacity (more than 500 tonnes). Some of the factors that should be considered in determining the appropriate size of a transfer facility include the capacity of collection vehicles, time required to unload, waste sorting, hours of station operation and time required to connect and disconnect the trailers from compactors.

Learning Activity 3.5

Convert 400 kg/m^3 to tonnes, i.e. $400/1000 = 0.4$ tonnes/m^3

(i) Number of services/vehicle load (N)

$$N = (C \times D)/W$$
$$= (12 \times 0.4)/2$$
$$= 2.4 \sim 2$$

(ii) Time required to collect the load

$$E = N \times L$$
$$= 2 \times 2$$
$$= 4 \text{ minutes}$$

Learning Activity 3.6

The solid waste collection in our ward is undertaken by a private contract system, and Bangalore Mahanagara Palike (BMP), which has appointed a Medical Health Officer, a Senior Health Inspector, a Junior Health Inspector and Sanitary

Daffedars, assists the private contractor. They have employed *pourakarmikas* for sweeping and collection of wastes from households and containers. The Health Inspectors are in charge of the muster roll. Workers assemble at 6:30 a.m. in the morning at a specified place. Each worker is given the equipment and the section of the ward, which he or she has to clean. Work ends at 10:30 a.m. and the workers assemble at the same place to give their muster roll. From 11 a.m. to 1:30 p.m. gang work is carried out in a particular area or even a whole ward depending on the need. The need may be a clogged drain or clearing of black spots. The attendance of each of the workers is recorded at the field by the inspectors. There is another muster roll at 1:30 p.m. At present records are not maintained on the quantity of wastes collected and their variation. Once in a while, the contractor of our ward is asked to determine the quantity of waste generated per day.

Learning Activity 3.7

The activities responsible for successful implementation of collection and transfer systems are as follows:

- Finalising and implementing the system management plan.
- Purchasing and management of equipment.
- Hiring and training personnel.
- Providing public information.
- Monitoring system cost and performance.

Learning Activity 3.8

(i) If the crew can service two customers in one minute, then in four hours (i.e., four × 60 minutes) the crew can service:
$X = 480$ customers per day.

(ii) The number of collection vehicles needed for a community is given by:

$$N = \frac{SF}{XW}$$

where N = number of collection vehicles required; S = total number of households serviced = 10,000; F = collection frequency, number of collections per week = 1; X = number of households a single truck can service per day = 4 and W = number of workers per week = 250.

Therefore $N = \frac{10000 \times 1}{250 \times 4}$, i.e., the corporation ward requires ten collection vehicles.

UNIT 4: WASTE DISPOSAL

Learning Activity 4.1

Based on a general observation, some of the constraints municipal authorities in our locality face with regard to effective waste disposal include poor political back up, inadequate infrastructure, insufficient funds and lack of public support. Let me explain these below:

(i) Waste disposal is as much a technical issue as a political one. For example, an elected representative, called a Corporator, administers a municipal ward, and is responsible for the general upliftment of the ward. But, generally, waste disposal is not considered a priority issue. However, our locality, and indeed, Bangalore is just awakening to the importance of waste disposal, and the Bangalore Agenda Task Force appointed by the State government has introduced a programme called the *Swachha Bangalore*. This programme involves door-to-door collection of solid waste. But, this is merely a temporary solution as the city requires an integrated waste management system. In the main, for any programme to be successful, sufficient political and governmental backing is necessary.

(ii) In our locality, importance is given to waste collection and not waste disposal. Due to the inadequate infrastructure, the municipality is unable to give sufficient importance to waste disposal.

(iii) Waste disposal is a costly process and sufficient funds are not available to the municipality to adopt latest techniques such as pyrolysis, gasification, etc.

(iv) Public support is essential for the success of any waste management programme. For example, people must stop dumping wastes indiscriminately on roadsides and in drains. Indiscriminate dumping makes waste collectors' work tedious and time consuming.

Learning Activity 4.2

Based on technical (i.e., composition of waste, existing practices and technology), social (i.e., health and income implication, and public opinions) and environmental (i.e., initial and long-term environmental risks) aspects, the best disposal option for our locality is composting. This is due to the fact that our locality is predominantly residential, generating mostly biodegradable wastes, and composting does not need long-term maintenance. Nor does it cause any adverse environmental effects.

Learning Activity 4.3

No, a sanitary landfill is not possible to manage wastes in our locality mainly because:

- it is not economically feasible for a small locality such as ours;
- composition of waste is mostly biodegradable;
- sufficient land is not available for setting up a landfill site.

Learning Activity 4.4

Typical landfill gases include methane, carbon dioxide, oxygen and nitrogen, and these adversely impact on public health and the environment. For example:

- Methane is flammable in air and can lead to unpredictable and uncontrolled subsidence and production of smoke and toxic fumes.
- Trace components such as aldehydes, alcohol and esters are toxicants when present in air at concentrations above occupational exposure standards.
- The landfill gases lead to global warming as these absorb reflected solar radiations.

It is vital that these emissions are controlled, and we can achieve this by controlling:

- waste inputs (i.e., restrict the amount of organic wastes);
- processes within the waste (i.e., minimise the moisture content to limit gas production);
- the migration process (i.e., put physical barriers or vents to remove the gas from the site and reduce gas pressure).

Since gas emissions cannot be easily prevented, removal by vents within the waste or stonewalled vents is the preferred option.

Learning Activity 4.5

The best way to control leachate is through prevention, which can be done with natural liners and synthetic liners.

- Natural liners, such as compacted clay or shale, bitumen or soil sediments, are less permeable, resistant to chemical attack and have good sorption properties.
- Synthetic (geo-membranes) liners such as high density or medium density polyethylene, are less permeable, easy to install and relatively strong and have good deformation characteristics.

Natural and synthetic liners can be combined to improve the efficiency of the containment system. The various treatments used to treat leachates are leachate recirculation, biological treatment (e.g., aerated lagoons, activated sludge process, rotating biological contactors, anaerobic treatments that include anaerobic filters, anaerobic lagoons and digesters) and physico-chemical treatment (e.g., flocculation-precipitation process).

Learning Activity 4.6

A sanitary landfill is essentially an engineered waste disposal option, where the environmental risk is controlled at an appropriate and acceptable level. But, if it is not properly managed, it can cause the following problems: (i) noise pollution due to vehicles such as earthmovers, compactors, etc., (ii) scavenger birds, vermins, insects, etc., are attracted to the landfill for breeding and they constitute potential health problems, (iii) leachates, if not treated or controlled, may pollute groundwater and surface water and (iv) gas released, due to degradation or volatilisation of waste components, causes problems such as odour, flammability and damage to health and the environment.

Uncontrolled disposal is a non-engineered waste disposal method in which wastes are dumped at a designated point without any environmental control, and it (i) causes odour problems due to the putrification of biodegradable wastes, (ii) provides a breeding place for disease vectors such as flies, mosquitoes, etc., and (iii) causes visual pollution, etc.

When properly managed, a landfill is a better and safer option of disposing wastes than uncontrolled dumping. Also, it is adaptable to all kinds of wastes and there is potential for by-product recovery.

Learning Activity 4.7

Landfill is a complex reactor where physical, chemical and biological processes transform polluting wastes into environmentally acceptable deposits. Due to the complexity of these processes and their potential environmental effects, monitoring is needed to confirm that the landfill works as expected. Knowledge of the chemical composition of leachate and gas is required, as leachates may contain toxic substances and, if not prevented, may contaminate the groundwater. Landfill gases that may be flammable, asphyxiating and noxious, pose a health hazard. A continued groundwater monitoring programme is essential for confirming the integrity of the liner system.

UNIT 5: WASTE PROCESSING TECHNIQUES

Learning Activity 5.1

Compaction of wastes is the method in which waste is densified so as to reduce its volume. This is done to improve the efficiency of collection and disposal of wastes. Compaction is done to increase the useful life of landfills and to reduce the quantity of material handled at the disposal site. It also brings down the cost involved in waste management.

Size reduction refers to the conversion of solid wastes into smaller portions. This helps to obtain the final product in reasonably uniform and considerably reduced size in comparison to the original form. It is important in the recovery of materials for reuse and for conversion to energy. In order to make a better fuel for incineration waste energy recovery facilities, size reduction is practised. It is also used prior to moisture reduction, drying and dewatering.

Learning Activity 5.2

1 tonne = 1000 kg

The weight fraction of the glass in the feed is given by the equation:

$$W_f = \frac{\text{Weight of sample}}{\text{Weight of material fed to the screen}} \times 100$$

$$= \frac{100 \times 1000 \times 0.08 \text{ kg}}{100 \times 1000 \text{ kg}}$$

$$= 0.08$$

Weight fraction of glass in screen underflow is given by:

$$Wu = \frac{\text{Weight of sample in underflow}}{\text{Total weight of material in underflow}}$$

$$= \frac{7.2 \times 1000 \text{ kg}}{10 \times 1000}$$

$$= 0.72$$

$$\text{Recovery (\%)} = \frac{UW_u}{FW_u}$$

Recovery efficiency is given by the equation:

$$= \frac{10 \times 1000 \times 0.72 \times 100}{100 \times 1000 \times 0.08}$$

$$= 90\%$$

Effectiveness is given by the equation:

$$Effectiveness = \text{recovery} \times \text{rejection}$$

$$= \frac{U \times W_u}{F \times W_f} \times \left[1 - \frac{U(1 - W_u)}{F(1 - W_f)}\right]$$

$$= \frac{(10 \times 1000)\ (0.72)}{(100 \times 1000 \times 0.08)} \times \left[1 - \frac{10 \times 1000\,(1 - 0.72)}{100 \times 1000\,(1 - 0.08)}\right]$$

$$= 0.87$$

Learning Activity 5.3

The heat required for drying can be applied by the following methods:

(i) Convection drying in which hot air is in direct contact with the wet solid waste stream.

(ii) Conduction drying in which wet solid waste stream is in contact with a heated surface.

(iii) Radiation drying in which heat is transmitted directly to the wet solid waste stream by radiation from the heated body.

UNIT 6: SOURCE REDUCTION, PRODUCT RECOVERY AND RECYCLING

Learning Activity 6.1

(i) Source reduction tops the hierarchy of solid waste management because of its potential to reduce system costs, prevent pollution, conserve resources and increase efficiency. Source reduction programmes are designed to reduce both the toxic constituents in products and quantities of waste generated. Source reduction is a front-end waste avoidance approach that includes strategies such as designing and manufacturing products and packaging with minimum volume and toxic content and with longer useful life.

(ii) Criteria to be considered in this regard are:

- Social and economic equity.
- Economic and administrative feasibility, efficiency and cost.
- Volume requirement and scarcity of materials and natural resources in product manufacture.
- Volume of product and its by-product that must be eventually disposed.

- Useful life, reusability and/or recyclability of a product.
- Priority of source reduction of more hazardous to less hazardous products.

Learning Activity 6.2

Recycling as such is not practised in our locality. However, waste buyers show interest in buying used newspapers, which considerably reduces the paper content in the waste generated. The waste buyers, in turn, sell the papers to retail waste dealers or wholesale dealers. Apart from this activity, the awareness and the advantages of recycling have not been given due importance, and the authorities concerned have not shown any interest in properly addressing recycling of waste in our locality.

Learning Activity 6.3

The factors that are involved in planning are:

- Building local expertise.
- Understanding and developing a recycling market.
- Fostering public education and involvement.
- Assessing local waste stream.
- Augmenting existing programme.
- Setting goals and objectives.
- Municipal coordination.
- Programme evaluation.

Learning Activity 6.4

Our locality does not have an organised recycling programme. However, informal sectors do contribute to recycling of wastes – paper, glass bottles, polyethylene envelopes, etc.

Learning Activity 6.5

The end uses for high-density polyethylene (HDPE) are non-food bottles, drums, toys, pipes, sheets and plastic pellets. The end uses of polyethylene terephthalate (PET) include plastic fibres, injection moulding, non-food grade containers and chemicals.

UNIT 7: RECOVERY OF BIOLOGICAL CONVERSION PRODUCTS: COMPOST AND BIOGAS

Learning Activity 7.1

Composting is the transformation of organic material (e.g., plant matter) through decomposition into a soil-like material called compost. Invertebrates (e.g., insects and earthworms) and microorganisms (e.g., bacteria and fungi) help in transforming the material into compost. Composting is a natural form of recycling, which continually occurs in nature. Decomposition occurs naturally wherever plants grow. Creating ideal conditions can encourage this natural decomposition. The microorganisms and invertebrates fundamental to the composting process require oxygen and water to successfully decompose the material. The end products of the process are soil-enriching compost, carbon dioxide, water and heat. For example, composting of organic wastes such as vegetable wastes produce an organic manure, which is good for garden plants.

Learning Activity 7.2

There are three processes of composting, viz., biological, chemical and physical. In biological processes, microorganisms such as bacteria, fungi and actinomycetes break down the organic matter present in the waste. As microorganisms begin to decompose the organic material, the carbon in it is converted to by-products like carbon dioxide and water and a humic end product called compost. Inoculum is added in case of a lack of microbial organisms. If all conditions are ideal for a given microbial population to perform at its maximum potential, composting will occur rapidly. The chemical environment is largely determined by the composition of material to be composted. Several factors determine the chemical environment for composting: carbon energy source, the correct amount of water, adequate oxygen, appropriate pH and the absence of toxic constituents that could inhibit microbial activity. A physical environment is essential for the composting process to proceed efficiently. This includes factors such as temperature, particle size, mixing and pile size. Smaller particle size is preferred, as it has more surface per unit weight, and therefore, facilitates more microbial activity, which leads to rapid decomposition. The optimum temperature range is between 32 and 60 °C. Temperature above 65 °C is not ideal for composting. Mixing is done to distribute moisture and air evenly and promote the breakdown of compost clumps.

Learning Activity 7.3

There are five basic stages involved in all composting practices, namely, preparation, digestion, curing, screening and finishing and storage or disposal. Preparation consists of sorting the recyclables, removal of non-combustibles, shredding, pulping, grinding and the addition of water sludge. There are five techniques of digestion, viz., windrow, aerated, static pile, in-vessel composting and anaerobic processing. These techniques vary in the method of air supply, temperature control, mixing/turning of the material, time required for composting and capital and operating costs. The amount of curing necessary at the initial stabilisation depends on the proposed use of the composted material. If the compost is to be pelletised and/or bagged, it must be completely stable before being finished. The amount of finishing required is also dependent on the final marketing of the compost. Regrinding and rescreening is done, so as to make the compost uniform in size. The efficiency of the composting process, i.e., the success or failure of the operation, depends on the method of disposal. Even where a good compost market exists, provisions must still be made for storage because the use of composting is seasonal, with greatest demand during the spring and fall.

Learning Activity 7.4

In the anaerobic composting process, facultative bacteria break down the organic materials in the absence of oxygen and produce methane and carbon dioxide. Single-stage digesters contain the entire process in one airtight container. The feedstock is first shredded, and before being placed in the container, water and possible nutrients are added to the previously shredded material. The single-stage digester may contain agitation equipment, which continuously stirs the liquefied material. As digestion progresses, a mixture of methane and carbon dioxide is produced. These gases are continuously removed from both first- and second-stage digesters and are either combusted on-site or directed to off-site gas consumers. A stabilised residue remains, when the digestion process is completed, which can be removed either by mechanical equipment or pumped out as a liquid. The digested residue may require further curing by windrow or static pile composting.

Learning Activity 7.5

Yes, I suggest biogasification of municipal solid waste. Biogasification is an effective way of recovery of waste in the form of biogas, which is a source of renewable energy. The benefits that can be accrued from biogas are production of energy, transformation of organic waste into high quality fertiliser, improvement of hygienic conditions through reduction of pathogens, worm eggs and flies, environmental advantages through protection of soil, water, air and woody vegetation, etc. Also, biogas is much more convenient to use than traditional fuels such as firewood, dried dung and even kerosene. Biogasification is particularly suitable for wet substrates, such as sludges or food waste, which present difficulties in composting.

Learning Activity 7.6

Although compost has been used for a variety of purposes, contaminants in the compost could be detrimental to the environment and public health as they affect the water (i.e., eutrophication and leaching to surface and groundwater), the land (i.e., acidic soil, nitrogen depletion in soil, impacts of acids, organic and metal contamination), the air (i.e., odour) and health (i.e., drinking of contaminated water, toxins in the finished product and pathogens).

UNIT 8: INCINERATION AND ENERGY RECOVERY

Learning Activity 8.1

Incineration is the controlled high temperature oxidation of primarily organic compounds to produce carbon dioxide and water. Incineration of municipal solid waste can reduce its volume by more than 90%. Incineration reduces financial expenses that are necessary to carry waste to the landfill site. Due to the reduction in haul distance, fuel cost is reduced. The thermal conversion products that can be derived from solid wastes include heat and gas, which can be recovered. The low level heat remaining in the gases after heat recovery can also be used to preheat the combustion air, boiler makeup water or solid waste fuel. Overall incineration increases the efficiency of solid waste management by producing less waste and reducing cost.

Learning Activity 8.2

Strategic long-term planning is necessary for developing a successful incineration facility. This includes:

- Facility ownership and operation (full service approach, merchant plants and turnkey approach).
- Energy market (marketing steam and marketing electricity).

- Facility siting (effect on residents, environmental impact, development plans, proximity to waste source, proximity to energy market, logistic concern and residual ash disposal).
- Facility sizing (waste supply, alternative waste management programme, waste stream characteristics, planning for facility disruption, facility financing and time frame).

Learning Activity 8.3

Mass-burning system has several advantages. Apart from being the widely technically proven incineration technology, a mass-burn facility can accept refuse that has undergone little preprocessing other than the removal of oversized items. This provides advantages such as convenience and flexibility. Due to the large facility size, the incineration unit can be used to recover greater quantities of steam and electricity. The steam generated is of higher quality, which is then passed through a turbine to produce electricity or through an extraction turbine to generate electricity as well as provide process steam for heating purposes. One of the disadvantages of mass-burn facility is that a local programme is needed to separate household hazardous wastes and recover certain materials (e.g., iron scrap) so as to help ensure environmentally viable incineration and resource conservation. Other disadvantages include increased cost of construction and land space.

Learning Activity 8.4

The three basic types of waste-to-energy (WTE) incineration involve the generation of electricity, steam or the cogeneration of both electricity and steam. Electricity is the most common form of energy produced and sold from WTE facilities. By directing the steam produced from a WTE system through turbine generator, electricity can be produced and sold. The second method uses steam generated from incineration facilities directly by a customer for manufacturing operations. This steam can be used to drive machinery such as compressors, for space heating and generating electricity. Cogeneration involves the combined production of steam and electricity. In cogeneration, high-pressure steam is used first to generate electricity; the steam leaving the turbine is used to serve the steam users.

Learning Activity 8.5

Electrostatic precipitators (ESP) are used for particle control. ESP uses electric forces to move the particle flowing out of the gas stream on to the collector electrodes. The particles get a negative charge, when they pass through an ionised field. The electric field forces the charged particles to the walls. The particles that are collected are removed by knocking them loose from the plates and by intermittent and continuous washing with water. The removal efficiency is 99%.

UNIT 9: HAZARDOUS WASTE: MANAGEMENT AND TREATMENT

Learning Activity 9.1

The major hazardous waste generated, though in small amounts, is that generated from garages and residences. There are four garages in operation in our locality and the major waste constituents that are considered hazardous include gas, fuel, antifreeze, brake fluid, motor oil, batteries, etc. Of these, the major one is motor oil. The two characteristics of motor oil are ignitability and toxicity. Motor oil readily causes fire and burns vigorously so as to create a hazard. The flash point varies in motor oil. The lower the flash point the greater the tendency for the oil to suffer vaporization loss at high temperatures and to burn off on hot cylinder walls and pistons. Toxicity of motor oil has a deleterious effect on aquatic animals and plants.

Learning Activity 9.2

Management of hazardous waste follows the functional elements of solid waste, viz., generation, storage and collection, transfer and transport, processing and disposal. Special facilities are used that have sufficient capacity to hold wastes accumulated over a period of several days. Small amounts are containerised, and limited quantities may be stored for periods covering months or years. The waste generator or a specialised hauler normally does the collection of hazardous wastes for delivery to a treatment or disposal facility. All storage containers remain unopened till they reach the treatment and disposal facility. Processing of waste can be done to recover useful materials and to prepare the wastes for disposal. Processing can be accomplished either on-site or off-site. Regardless of their forms, most hazardous wastes are disposed off either near the surface or by deep burial. When landfills are used for such wastes, precautions must be taken to avoid the rupturing of containers during the unloading operation and the placement of incompatible wastes in the same location. Also provisions should be made to prevent any leachate from escaping from landfills, which can be done both visually and by monitoring wells.

Learning Activity 9.3

The factors required for land treatment operation involve the consideration of waste and soil characteristics, microorganisms and waste degradation.

Learning Activity 9.4

Waste minimisation means reduction, to the extent feasible, of hazardous waste that is generated prior to treatment, storage and disposal. The advantages include reduction of the total volume and toxicity of hazardous waste or reduction in both. By using materials carefully so as to reduce the generation of waste, pollution is reduced, resources are conserved and costs of waste disposal are minimised. Pollution prevention is the use of materials, processes or practices that reduce or eliminate the creation of pollutants or wastes at the source. The advantages include reduction of the use of hazardous and non-hazardous materials, energy, water or other resources as well as those that protect natural resources through conservation which is more efficient.

UNIT 10: INTEGRATED WASTE MANAGEMENT (IWM)

Learning Activity 10.1

Waste is an inevitable product of society. Historically, health and safety have been the major concerns in waste management, and solid waste management practices were initially developed to address these concerns, which were the result of indiscriminate dumping of waste without proper collection and disposal. Managing this waste is now a need that society has to address. In addition to ensuring human health and safety, a sustainable system for solid waste management systems must be environmentally effective, economically affordable and socially acceptable. An integrated approach to solid waste management can deliver both environmental and economic sustainability. An integrated system would include an optimised waste collection system and efficient sorting followed by options such as material recycling, biological treatment, thermal treatment and landfill.

Learning Activity 10.2

The steps involved in implementing IWM are planning, price, publicity, political support and perseverance. A well-devised and comprehensive plan is necessary to implement a successful programme. Planning should involve political bodies, waste generators, haulers, construction contractors, etc., as each group has the potential to strengthen or derail a project. Sound economic analysis that considers the resources of the community and adverse impacts on the environment is necessary to adopt alternative strategies for managing waste. A plan for informing the public about the programme's progress should be developed and implemented as the programme succeeds. The support for a programme can erode quickly, unless attention is paid to keeping the programme on the public agenda and maintaining strong and positive public support. Political support is often crucial to obtaining finance and ensuring that the programme gets the resources needed to construct facilities and operate them efficiently. A community waste management programme must be prepared for the long term. The ultimate key to success is the political and public will to persevere until the programme is in place.

Learning Activity 10.3

The stages are goal and scope definition, life cycle inventory analysis (LCI), life cycle impact assessment (LCIA) and life cycle interpretation.

Learning Activity 10.4

The stages are concern, involvement, issue resolution, alternatives and consequences.

Distance Learning in Environment Engineering

Aimed at providing in-service professionals with the knowledge, skills and attitudes that enable them to recognise and solve problems in their domains of study and future work, Indian Institute of Science (IISc.) offers distance learning courses in Environmental Engineering.

The Institute uses various media for course delivery. The printed course materials, for example, are self-learning in nature, simulating learner-teacher interaction. Furthermore, appropriate information and communication technologies are used to facilitate just-in-time contacts among distance learners and teachers. The delivery model also incorporates an instructional system that uses performance-based learning processes and this enables you to demonstrate your understanding of the content-areas. The intensive competence-based course format allows you to take more responsibility for your own learning. In order to cater to the varying learning styles, the Institute offers the subject material in the following modes:

(i) Downloadable mode: Downloadable mode empowers to offer education in a dynamic form to students. Students can access lessons, assignments and submit their scripts online. They need not brood if they miss lessons - they can download their lesson notes and access archived lessons, tutorial sessions, lectures, etc.

(ii) Read-only mode: Students will be able to only read the course concepts, downloads or interactive sessions are unavailable.

(iii) Classroom interaction: The interactive learning provides an environment for students to talk and discuss freely about any related topics on or off line.

- Synchronous and asynchronous communication (X-chat, discussion forum, bulletin board, email, etc.): Client Server Model that works with Internet Relay Chat Server Protocol and allows servers to connect to each other effectively forming a network. Protocol is used by servers to talk to one another. Instructor and participants can talk with each other and discuss on related topics for a stipulated time given by the instructor. In mailing-group, participants can become a member of Mailing-group by registering their names in their group. Once approved by the instructor, participants can post their questions to this group ID. This way the course instructor or any other participant can respond to questions that will be read by all the members in that group including the instructor.

(iv) Practice orientation: Participants progress through the curriculum by demonstrating their competence in a variety of skill and knowledge areas. Student who demonstrates a competency is showing the ability "to do" something to put knowledge and principles into practice. After completing all examinations, students are required to apply the course content in his or her work setting and submit an example of how they achieved this goal.

Currently, the following two courses are on offer during January–May and August–December Sessions at the Indian Institute of Science, Centre for Continuing Education (*http://cce.iisc.ernet.in*):

1. **Environmental Management**
2. **Municipal Solid Waste Management**

Note that the course home page is the central location for course information. The faculty-in-charge of the course(s) will post their syllabi, grading policies, assignments, course schedules, etc., on the page. This apart, during the briefing session, you will

- learn about email functions,
- learn to use the Chat Room; an exciting facility that helps you communicate directly with your peers and the faculty at a distance,
- learn to communicate with your faculty in real-time,
- learn to view course grades in the Grade book,
- learn to use the Drop-box facility to submit and retrieve assignments,
- learn to make one-on-one communication between you and your faculty in connection with assignments,
- find out about the Document sharing, a public place in which you can upload documents to share and conduct "workshop",
- learn about the Online Help button for detailed, context sensitive instructions on how to use the tools.

Indian Institute of Science

(http://www.iisc.ernet.in)

The Indian Institute of Science (IISc) was started in 1909 through the pioneering vision of J.N. Tata. Since then, it has grown into a premier institution of research and advanced instruction, with more than 2000 active researchers working in almost all frontier areas of science and technology. IISc is an institute of higher learning and is constantly in pursuit of excellence. It is one of the oldest and the finest centres of its kind in India and has a very high international standing in the academic world as well.

The Institute currently has more than forty academic departments, pursuing research in different areas of science and engineering. The departments belong to one of the two faculties of the Institute, viz., the Faculty of Science and the Faculty of Engineering. The Faculty of Science awards PhD degrees through the Regular Research and Integrated PhD programme, while the Faculty of Engineering awards ME, MTech and MDes degrees in addition to PhD and MSc (Engg).

The Institute houses a huge collection of journals and textbooks in the J.R.D Tata Memorial Library. The Library also subscribes to many journals and publications in the electronic form. The Supercomputer Education and Research Centre makes available state-of-the-art computational facilities to the staff and students. The Centre runs round-the-clock, seven days a week and 365 days a year, helping the researchers to break many barriers in computational science and engineering.

The Institute has pioneered advanced education in India and has been making many significant contributions to frontier areas of research. The number of students in the Institute is kept deliberately small in order to focus on quality. It has been able to innovate and introduce (a) new systems of imparting knowledge and (b) educational reforms such as offering courses under unit system. The Institute was the first to introduce (i) Masters' programmes in engineering, (ii) the Integrated PhD Programmes in Biological, Chemical, Physical and Mathematical Sciences for science graduates, (iii) the new IISc Young Fellowship programme for the first 20 rank holders at the + 2 level and (iv) IISc Young Engineering Fellowship programme for merited III year BE/ BTech students.

The Institute has been primarily responsible for starting many fields of activity that include (i) Aerospace, Automation, Chemical, Civil, Computers, Electronics, Electrical, Mechanical, Metallurgical and Telecommunication Engineering in

Engineering and (ii) Biochemistry, Biophysics, Ecology, Materials Science and Solid State & Structural Chemistry in Science. It has also acted as a reservoir from which leadership and human resources for the future scientific developments and industries can be drawn. It is now involved in several emerging areas of national importance: space science and technology, environmental and atmospheric sciences, endocrinology, genetic engineering, rural technology and energy problems.

The Institute has a Centre for Scientific and Industrial Consultancy, through which significant R&D work is done on identified projects sponsored by industries. The know-how so generated has been transferred to industries. Transfer of technology has also taken place in areas such as low-cost housing and renewable sources of energy, which benefit the society. Similarly, the facilities available at the Institute such as low and high speed wind-tunnels, water-tunnel, high-speed computers and sophisticated instruments have been helping public and private sector industries, and defence. There has also been a certain amount of social utilisation of work in biosciences. For instance, (i) plant tissue culture of sandal wood, eucalyptus and teak wood, (ii) disease control in silk worms and (iii) nutritional value enhancement of rice strains. Some recent projects include the development of cryogenic equipment and vessels for advanced aircraft, fracture analysis of industrial and space launch vehicle components, thermo-metallurgical modelling of steel industries, acoustic absorbers to reduce acoustic pollution in industrial environment and automobiles and microbial techniques for gold extraction from mines. In 1997, the Institute released technologies relating to AIDS diagnostic kit, biomass gasifier and ferroelectric materials.

In all these endeavours, the Institute, with an awareness of its noble tradition and the need for maintaining a high quality in all its activities, strives to contribute to the scientific, academic and technological goals of India. Academic programmes leading to ME or MTech by course work and PhD and MSc (Engg.) by research are available in almost all the departments. A unique feature of the ME/MTech programmes at the Institute is the credit system which allows a student to tailor the Courses to suit his/her aptitude and interest. Every ME/MTech programme has a set of hard core Courses specified as an essential requirement whereas the student can take rest of the credits from many Courses available in his/her parent or other departments and also do a dissertation work on the topic of his/her choice. These Courses attract highly motivated, accomplished students, in addition to several sponsored candidates from R&D labs./industries and also from educational institutions under the Quality Improvement Programme (QIP).

A flexible approach to the ME/MTech programme has been recently introduced in a few departments, which enables candidates from industry to take Courses leading to ME/MTech, spread over a longer period than the regular programme, without having to leave their workplaces. Research programmes leading to MSc (Engg.)/PhD degrees are the main thrust in many departments. The programme has a limited amount of course work, essentially to prepare the student to carry out the research, but the main emphasis is on the thesis work. Excellent laboratory and computational facilities, a unique library, outstanding faculty with strong interests in both basic and applied sciences, make "dream come true" kind of environment to students with strong interest and aptitude for research. These programmes attract students from all corners of the country and beyond.

The External Registration program leading to MSc (Engg)/PhD provides a most attractive mechanism through which candidates from industry/R&D can work with the faculty at the Institute under the joint guidance of Institute faculty and senior officers, scientists in their parent organisation and acquire a research degree of the Institute with a minimal residential requirement. This has been one of the most successful programmes with nearly 30% research student enrolment in the Engineering faculty.

The Faculty being highly qualified and exposed in specialised areas of national importance attract a great deal of sponsored research and have the opportunity to participate at high-level national and international science and engineering meetings. This feature enables the faculty to respond rapidly to the changing scenario in science and technology and give initiations by introducing and nurturing new subjects. Thus, every research effort becomes indeed a step in a new direction, providing enriching experience of learning/discovering in the ever-changing scene of engineering and science. During the 1997-98 session, the Institute introduced three new Courses viz. (1) ME Programme in Microelectronics, (2) ME Programme in Signal Processing and (3) MDes Programme in Product Engineering and Design. To keep up with the changing scenario of science and technology development in the country, more Masters' Degree Courses are introduced since the 1999-2000 session, viz. (1) the 2-year MBA programme (which replaces the existing MTech programme) in the department of Management Studies and (2) the 2-year MTech (Computational Science) programme in the Supercomputer Education and Research Centre.

The Indian Institute of Science has for a long time been concerned with the ways to make its expertise and facilities available to qualified technical personnel in industries, universities and research establishments. On a small scale, the Institute's facilities are always being used through short-term/refresher courses, but these activities are not coordinated by a single entry. The Centre for Continuing Education (CCE) has therefore been set up in 1975-76 to function as a focus for all such activities and to promote a variety of programmes relevant to specific target groups (*http://cce.iisc.ernet.in*).

The Energy and Wetlands Research Group at the Centre for Ecological Sciences (CES), is actively involved in studies and training on issues related to environment, water resources, energy, ecology, wetlands, geographic information systems, environmental impact assessment and natural resource management. The information about research activities is available at *http://ces.iisc.ernet.in/energy/Welcome.html*

Recognizing the need for continuing professional development, the Commonwealth of Learning (COL) provided five high quality environmental engineering texts earlier developed through a UNESCO-COL partnership to the Energy and Wetlands Research Group at the Centre for Ecological Sciences (CES), Indian Institute of Science (IISc.) They jointly examined the relevance of, and the degree of modification required for, offering at a distance those texts for practising professionals in environmental engineering.

In collaboration with the Karnataka Environment Research Foundation, Indian Institute of Technology (IIT, Mumbai), National Environmental Engineering Institute (NEERI) with the financial support from the Commonwealth of Learning (COL), the Energy and Wetlands Research Group at the Centre for Ecological Sciences has developed five Environmental Engineering Courses in self-learning format suitable for continuous professional development. Based on the instructional problems, the audience analysis, instructional goals and objectives, and an understanding of the desired Course content, an outline of the content to be covered in the environmental engineering Courses were developed. Five Courses were developed simultaneously involving authors with respective domain expertise. The Courses developed in self-learning materials (SLM) for continuous professional development of engineers and the institutions are:

1. Municipal Solid Waste Management
2. Ground Water and Soil Pollution from Agricultural Activities
3. Air Pollution Control
4. Municipal Water and Waste Water Treatment
5. Environmental Management

These Courses are directed at practising professionals (in service professionals), engineers and technologists, in India, South Asia and the developing countries of the Commonwealth, who are all involved in environmental management in some form. The treatment, therefore, is not deeply technical or mathematical. In other words, building on their existing repertoire of skills base, through these Courses the target groups will be exposed to the new knowledge that is available in the field.

Next, existing materials along with outline of the content were reviewed by a programme review committee (at one week duration round table meeting) considering the relevance for distant learners with widely varied and non-traditional experiential backgrounds. The programme review committee chaired by COL included content experts, authors of various Courses, instructional design expert, project manager and a programme coordinator.

Instructional design was done based on content, need for instruction, audience need assessment, etc. Content was organised and developed using relevant examples from India that relate the content to a context understood by the students. This problem was addressed through discussions with a sampling of the target audience. The review of content and design was done at six months interval (mid term review and pre final review).

Evaluation was done including both formative and summation approaches to determine if the instructional methods and materials are accomplishing the established goals and objectives.

- Formative evaluation focussed on course strengths and weaknesses, technical or delivery concerns, and content areas in need of further coverage.
- Summative evaluation was done through validation and pilot testing. Pilot testing included both brainstorm ways and open discussion of Course contents. This provided a data for Course revision and future planning.

Within the context of formative and summative evaluation, data are collected through quantitative and qualitative methods combining quantitative measurement of student performance with open-ended interviewing and non-participant observation to collect and assess information about attitudes toward the Course's effectiveness and the delivery technology. Careful analysis of pilot testing results helped in identifying gaps or weaknesses as well as strengths and successes in the instructional process.

Revision plans typically were a direct result of the evaluation process in tandem with feedback from validation exercises involving Course authors from and content specialists. Also, it was evident from the pilot testing exercises that the instructor's own reflection on Course strengths and weaknesses were the best source of revision ideas. Most revisions were minor, such as breaking a large and unwieldy instructional unit into more manageable components, increasing assignment feedback, or improving student-to-student interaction. In one Course, major revisions were implemented.

The development of delivery models required integration of print, CD, and internet technology in concert with face-to-face communication. The challenge here was to integrate delivery components, based on identifiable learner needs, content requirements, and technical constraints.

Distance Education System and Study Skills

Studies across the globe, comparing distance education to traditional classroom instruction, indicate that teaching and learning at a distance are as effective as traditional instruction, when the learning materials, method and technologies used are appropriate to the instructional tasks and when student-to-student interaction and teacher-student feedback are robust. The typical challenges posed by distance education are countered by opportunities to:

- reach a wider student audience,
- meet the needs of students who are unable to attend on-campus classes,
- involve outside speakers who would otherwise be unavailable, and
- link students from different social, cultural, economic, and experiential backgrounds.

These days with advances in science and technology a wide range of technological options are available to the distance educator. They can be grouped into four major categories namely voice, video, print and web (Internet) options. Voice tools include passive or one-way technologies (tapes, radio, etc.) and the interactive technologies (telephone, audio conferencing, etc.). Video tools include still images such as slides, pre-produced moving images (e.g., film, videotape), and real-time moving images combined with audio-conferencing (one-way or two-way video with two-way audio). Print version in self-learning format is a foundational element of distance education programmes and the basis from which all other delivery systems have evolved. Various print formats that could be used for distance learning are: textbooks, study guides, workbooks, course syllabi, and case studies. With the advance and affordability of communication technology, web or Internet based learning become a reality. Technology plays a key role in the delivery of distance education. However, the focus of the programme is on the instructional outcomes rather than on the technology of delivery. In other words, the main thrust is on the needs of the learners, the requirements of the content and the constraints faced by the teacher, before selecting a delivery system. Typically, this systematic approach has resulted in a mix of media, each serving a specific purpose as touched upon below:

- Print component provides much of the basic learning content in the form of learning materials (which are in self-learning format), supplemented by suggested readings and other support materials in print.

- Interactive sessions provide real-time/face-to-face (or voice-to-voice) interaction (audio or video conferencing). This has proved to be, among others, a cost-effective way to incorporate guest speakers and content experts.
- Computer conferencing (e.g., chat session, discussion forum, bulletin board) or electronic mail to send messages, assignment feedback, and other targeted communication to one or more class members is also prevalent. This helps to increase interaction among students and between students and teachers.
- Pre-recorded video tapes/CD/DVDs are used to present class lectures and visually oriented content.
- Fascimile services are used, where affordable and required, to distribute assignments, just-in-time announcements, receive student assignments, provide timely feedback, etc.

Using this integrated approach, the task is to carefully select among the technological options depending on students (infrastructural facilities available in their locality), subject requirements (field visit, case studies, etc.) and institutional commitments. The goal is to build a hybrid instructional media, meeting the needs of the learner in a manner that is instructionally effective and economically prudent.

Distance education programmes have been initiated with careful planning and a focussed understanding of the Course requirements and student needs. Appropriate technologies have evolved through the efforts of many individuals and organisations, which rely on the consistent and integrated efforts of students, faculty, facilitators, support staff, and administrators. When instruction is delivered at a distance, additional challenges result because students are often separated from others sharing their backgrounds and interests, have few opportunities to interact with teachers outside the class, and rely on technical linkages to bridge the gap separating class participants. Hence, the success of distance education effort rests squarely on the shoulders of the faculty and resource persons. In a traditional classroom setting, the instructor's responsibility includes assembling Course content and developing an understanding of student needs.

NEED FOR ENVIRONMENTAL ENGINEERING DISTANCE LEARNING COURSES

Rapid industrialisation of developing countries has led to an increasing demand for continuing professional development for environmental engineering personnel at all sectors, including industrial and infrastructure management sectors. Recognizing the need for continuing professional development, the Commonwealth of Learning (COL) provided five high quality environmental engineering texts earlier developed through a UNESCO-COL partnership to the Indian Institute of Science (IISc.). They jointly examined the relevance of, and the degree of modification required for offering at a distance, those texts for practising professionals in environmental engineering. This resulted in a COL-funded project with IISc., taking the responsibility of coordinating and contributing to the design and development of a series of learning materials in Engineering Sciences for open distance learning.

Based on the analysis of target audience, learning goals and objectives, and the desired course content with particular relevance to India, an outline of the content was created, which was reviewed by a committee of content and instructional design experts. The development of five self-learning environmental engineering Courses included formative evaluation by the project review committee and electronic list discussions. The summative evaluation included validation and pilot testing and brainstorm sessions to ascertain the suitability of materials to meet the project goals and objectives. The materials thus produced were revised on the basis of the feedback and are on offer at a distance currently within India. Nonetheless, these Courses when suitably adopted with contextual examples can meet the continuous professional development requirements of the developing countries in the Commonwealth. It is also believed that these Courses can be appropriately modularised to contribute to the proposal put forward by the World Federation of Engineering Organisations for the World Engineering Partnership for Sustainable Development.

Course Content

COURSE 1 : MUNICIPAL SOLID WASTE MANAGEMENT

1. Municipal Solid Waste Management: An Introduction
2. Generation and Characteristics of Waste
3. Waste Collection, Storage and Transport
4. Waste Disposal
5. Waste Processing Techniques
6. Source Reduction, Product Recovery and Recycling
7. Recovery of Biological Conversion Products: Composts and Biogas
8. Incineration and Energy Recovery
9. Hazardous Waste: Management and Treatment
10. Integrated Waste Management (IWM)

Due to rapid increase in the production and consumption processes, societies generate as well as reject solid materials regularly from various sectors—agricultural, commercial, domestic, industrial and institutional. The considerable volume of wastes thus generated and rejected is called solid wastes. In other words, solid wastes are the wastes arising from human and animal activities that are normally solid and are discarded as useless or unwanted. This inevitably places an enormous strain on natural resources and seriously undermines efficient and sustainable development. One of the ways to salvage the situation is through efficient management of solid wastes, and this is the focus of this Course, Management of Municipal Solid Waste. In the ten Units that constitute this Course, we will discuss the processes involved in the management of solid wastes—from waste generation to final disposal.

In Unit 1, we will describe solid wastes and introduce you to the classification of solid wastes and the functional elements, such as waste generation, storage, collection, transport, processing, recovery and disposal, in the management of solid wastes. In Units 2 to 7, we will explain with the support of case studies each of these functional elements. In Unit 8, we will explain the treatment of solid wastes by incineration and energy recovery from the incineration process. Subsequently, in Unit 9, we will deal with the treatment and management of hazardous (biomedical) wastes. Finally, in Unit 10, we will discuss the concept of integrated waste management.

COURSE 2: ENVIRONMENTAL MANAGEMENT

1. Principles of Environmental Management
2. Principles of Ecology, Environment & Environmental Management
3. Policies and Legal Aspect of Environmental Management
4. Overview of Environmental Impact Assessment (EIA)
5. Environmental Audit
6. Life Cycle Assessment as EM Tool
7. Environmental Management Systems Standards: ISO 14000 (EMS)
8. Related Issues in Environmental Management
9. Environmental Design
10. Environmental Economics

Due to unplanned developmental activities as well as ever-increasing population, which has caused enormous strain on the environmental resources, societies across the world face several problems of environmental degradation. However, it is imperative to maintain a balance between the capacity of the environment and the quantum of sustainable utilisation. This is only possible by understanding the environment in its totality and the principles of its scientific management.

Environmental management (EM) has become one of the most used terms in recent times. But, what exactly does the term mean and entail? Different individuals belonging to different disciplines approach EM differently and therefore it is difficult to find a single, comprehensive view of EM. However, we will discuss in detail the various issues relating to EM in the ten Units that constitute this Course.

We will first discuss the fundamentals of EM and ecosystem in Unit 1 and, subsequently, the various environmental policies, legislations and international treaties in Unit 2. In Units 3 and 4, we will deal with the concept of environmental impact assessment (EIA) and the preparation of EIA report, respectively. We will discuss the methodology and processes of environmental auditing in Unit 5. In Unit 6, we will introduce you to life cycle assessment (LCA) in the context of EM. In Unit 7, we will explain the various EM system standards. We will then take up the issues and techniques relating to EM in Unit 8. Finally, we discuss environmental design and economics in Units 9 and 10, respectively.

COURSE 3: AIR POLLUTION CONTROL

1. Introduction to Air Pollution
2. Air Pollution: Impact, Legislation and Standards
3. Air Pollution Measurement

4. Prevention of Air Pollution
5. Air Pollution Meteorology
6. Air Pollution Modelling
7. Air Pollution Control: Particulate Matter I
8. Air Pollution Control: Particulate Matter II
9. Air Pollution Control: Volatile Organic Compound (VOCs)
10. Air Pollution Control: Sulphur Dioxide and Oxides of Nitrogen

A healthy person inhales about 16 kg of air per day. If the air is polluted, it will have serious implications on the health. Inhalation of polluted air considerably reduces our intake of oxygen, adversely affecting our health. However, on the one hand, we treat the atmosphere as a vast sink for air pollutants, and, on the other, we take clean air for granted and assume that it is a free and unlimited resource.

According to a World Bank study (2000), the number of premature deaths due to air pollution in India has increased by almost 30%. One of the reasons for this increase, the study reveals, is indoor air pollution. Air pollutants released indoors are likely to be more harmful to health because people are close to the place of combustion. In fact, indoor air pollution, caused by the combustion of non-commercial sources of fuels such as bio-fuels in homes that are poorly ventilated, kills thousands of young children and women annually. Put differently, children, women and the disabled staying indoor are subjected to pollution levels that may go up to 20 times the acceptable limits. Note that while almost three-fourth of Indian households use fuels such as firewood, crop leftovers, cow dung and other organic wastes, the remainder uses coal, mineral oil and compressed gases.

Note, also that as air pollutants get dispersed throughout the atmosphere, air pollution is not merely a local problem at the villages, large urban centres or industrialised regions. It, indeed, is a global problem.

Now, are there ways by which we control/prevent air pollution and ensure the present and future generations have abundant supply of clear air?

In this Course, Air Pollution, we will discuss the important issues pertaining to air pollution and its control/prevention. In the ten Units that constitute this Course, we will deal with the various sources of air pollution and the environmental and health impacts as well as the appropriate measures to control/prevent air pollution. We will begin the Course by discussing the sources and classification of air pollutants in Unit 1. Subsequently, in Unit 2, we will study the impacts of air pollution at various levels – personal, local, regional, continental and global – and environmental legislations and air quality standards. While Unit 3 considers air pollution measurement, Unit 4 discusses air pollution prevention through process, raw material and operational changes that can be brought about in process industries and other activities. We will then explain the meteorological aspects of air pollution in Unit 5 and list the different air pollution models and their application in Unit 6. In Units 7 and 8, we will discuss the different devices for the separation of particles from polluted gases. In Unit 9, we will consider air pollution control for volatile organic compounds (VOCs) in gases. In Unit 10, we will discuss the different processes and operations for the removal of sulphur dioxide and oxides of nitrogen from the polluted gases.

COURSE 4: SOIL AND GROUNDWATER POLLUTION FROM AGRICULTURAL ACTIVITIES

1. Agricultural Activities and Environmental Impacts
2. Soil: Process and Properties
3. Groundwater Hydrology
4. Soil and Groundwater Pollution
5. Irrigation and Drainage: Impact and Remedial Measures
6. Environmental Impacts of Fertilisers
7. Pesticides and Environmental Impacts
8. Monitoring of Soil and Groundwater Pollution
9. Soil and Groundwater Treatment
10. Conservation and Management of Soil and Groundwater

Today, environmental pollution is a major concern all over the world. When sources of pollution are enumerated, agriculture is, with increasing frequency, listed as a major contributor. Agricultural activities can contribute to deterioration in soil and water quality through the release of several pollutants such as pesticides, animal manure, fertilisers, sediments (due to siltation), and other sources of organic and inorganic matter. Hence, in this Course, "Soil and Groundwater Pollution from Agricultural Activities", we will introduce you to major agricultural activities in India and their impact on soil and groundwater. In addition to this, you will be introduced to various remedial measures. In Unit 1, you will learn the basic aspects of agricultural activities. Subsequently, we will introduce you to the soil properties, classification and processes in Unit 2 and in Unit 3 you will learn groundwater characteristics, movement and recharge aspects. This is followed by discussion on soil and groundwater pollution from various sources, impacts of irrigation, drainage, fertiliser and pesticide on soil and groundwater in Units 4 to 7. In the eighth and ninth Units, various sensors/devices used to monitor pollution in soil and groundwater with their remedial measures will be introduced. Finally, conservation and management of groundwater and soil are presented in tenth Unit.

COURSE 5: MUNICIPAL WATER AND WASTEWATER TREATMENT

1. Water as a Resource: Some Related Issues
2. Municipal Waste Water Generation, Conveyance and Disposal
3. Water Treatment Process
4. Wastewater Treatment: Physical and Chemical Processes
5. Wastewater Treatment: Biological Methods
6. Wastewater Reuse
7. Sludge and Septage: Treatment and Disposal
8. Wastewater Disposal into Natural Water Bodies
9. Wastewater Disposal on Land – Land Application
10. Selection, Planning and Implementation Guideline for Water and Wastewater Treatment Systems.

In this Course, "Municipal Water and Wastewater Treatment" we will learn the concepts of water and wastewater treatment. In Unit 1, we will discuss concepts of water availability, quality, attributes, use-based classification, hydrological cycle, etc. In the latter part of the Unit, water quality for various uses, municipal water supply complexities, water leakage have been explained. Unit 2 on the other hand discusses aspects related to municipal wastewater collection, conveyance and treatment. Various components of wastewater collection system have been explained along with concepts such as type of pollutants, mass balance, nitrification, reaeration and dosage. Water supply and treatment processes have been described in Unit 3. Physical and chemical processes involving screening, sedimentation, flocculation, filtration, chemical precipitation, softening and disinfection have also been explained. Unit 4 pertains to physical and chemical treatment processes including ammonia stripping, foam fractionation, reverse osmosis, ion exchange, electrochemical treatment, electro- dialysis etc.

Unit 5 details the biological treatment processes for wastewater. It is important to understand the concepts of biological growth, growth kinetics for aerobic and anaerobic systems as well as practical aspects of biological treatment. One of the important objectives of wastewater treatment is to meet the regulatory requirement; however, in recent times, disposal of treated wastewater is not viewed as an environment friendly option. There is a lot of emphasis on investigating the alternative use of the wastewater for municipal, industrial, agricultural, forest or for recreational purposes. We will discuss these aspects in Unit 6. The treatment processes always produce sludges of different qualities and quantities that need appropriate treatment and disposal, which is discussed in Unit 7. Concepts and standards relating to discharge and receiving water quality related to treated wastewater disposal in lakes, rivers, estuaries and oceans are discussed in Unit 8.

Wastewater application and disposal on land have been discussed in Unit 9. Finally, we will describe in Unit 10 the selection, planning and implementation guidelines for water and wastewater treatment processes.

COURSE SCHEDULE

I. Classroom briefing and introductory sessions at the beginning of the course (2–3 days).
II. Interactive session through Web and Email for three months.
III. Mid-term exam after the third month.
IV. Field Work in the fourth month and final contact sessions for three days.
V. Final exam at the end of the fourth month.

QUALIFICATIONS AND REQUIREMENTS

This Course is open to all graduates. Participants are expected to take part in all activities that form a part of the Course; in addition, the participants should work on individual projects. During the Course, participants should compulsorily attend interactive chat sessions, for which some grading will be given. Participants should actively participate in the sessions for best results. A certificate along with grades will be given to all the successful participants.

EVALUATION

Evaluation will be done at two levels, namely, the concept level and the module level. At the concept level, after student goes through each concept, he/she is evaluated online by randomly displaying multiple-choice questions. Each question carries some weight or marks. Based on his/her test performance, the system checks the number of questions correctly answered and the performance is evaluated. At the module level, after the students complete all the concepts, they are given the final exam, which will be held at the Centre for Ecological Studies (CES)/ Centre for Continuing Education. The concept level evaluation marks will also be considered for grading. Completion of a project is a partial requirement for the Course.

Index